WATER AND FERTIGATION MANAGEMENT IN MICRO IRRIGATION

Research Advances in Sustainable Micro Irrigation

VOLUME 9

WATER AND FERTIGATION MANAGEMENT IN MICRO IRRIGATION

Edited by
Megh R. Goyal, PhD, PE

Apple Academic Press Inc. | Apple Academic Press Inc.
3333 Mistwell Crescent | 9 Spinnaker Way
Oakville, ON L6L 0A2 | Waretown, NJ 08758
Canada | USA

© 2016 by Apple Academic Press, Inc.

First issued in paperback 2021

Exclusive worldwide distribution by CRC Press, a member of Taylor & Francis Group

No claim to original U.S. Government works

ISBN-13: 978-1-77463-380-9 (pbk)
ISBN-13: 978-1-77188-106-7 (hbk)

Library and Archives Canada Cataloguing in Publication

Water and fertigation management in micro irrigation/edited by Megh R. Goyal, PhD, PE.

(Research advances in sustainable micro irrigation ; volume 9)
Includes bibliographical references and index.
ISBN 978-1-77188-106-7 (bound)
1. Microirrigation. 2. Water-supply--Management. 3. Fertilizers--Application--Management.
I. Goyal, Megh Raj, editor II. Series: Research dvances in sustainable micro irrigation; v.9

S619.T74W28 2015   631.5'87   C2015-903928-2

Library of Congress Cataloging-in-Publication Data

Water and fertigation management in micro irrigation / [edited by] Megh R. Goyal, PhD, PE.

pages cm. -- (Research advances in sustainable micro irrigation ; volume 9)
Includes bibliographical references and index.
ISBN 978-1-77188-106-7 (alk. paper)
1. Microirrigation. 2. Irrigation farming. 3. Water in agriculture. 4. Soils--Fertilizer move-ment. 5. Arid regions agriculture I. Goyal, Megh Raj.

S619.T74W38 2015   631.5'87--dc23   2015022927

Apple Academic Press also publishes its books in a variety of electronic formats. Some content that appears in print may not be available in electronic format. For information about Apple Academic Press products, visit our website at **www.appleacademicpress.com** and the CRC Press website at **www.crcpress.com**

CONTENTS

LIST OF CONTRIBUTORS

E. E. Abd-El Aaty, PhD
Agricultural Engineer, Soil Conservation, Desert Research Center, Ministry of Agriculture and Land Reclamation, 1 Mathaf El-Mataria Street, El-Mataria, Cairo-Egypt B.O.P. 11753. Tel.: +202 26332846, Mobile: +2 01063031920 Fax: +202 26357858

M. Kh. Afifi, PhD
Professor at Agric. Eng. Dep., Fac. Agric. Zagazig University

Abdurrahman Ali Alazba, PhD
Professor in Water Research, College of Agriculture and Food Sciences, Kingdom of Saudi Arabia, PO Box 2460/11451, Riyadh, Saudi Arabia

Natarajan Asokaraja, PhD
Professor and Head, Dept. of Agronomy, TNAU, Coimbatore. Mobile: +91 9367789899. E-mail: nasokaraja@yahoo.com

Timothy Coolong
Associate Professor in Vegetable Production, Department of Horticulture, Tifton Campus, University of Georgia, 2360 Rainwater Road (mailing), 4604 Research Way, Tifton, GA 31793 (shipping), Tifton, GA 31793–5766. E-mail: tcoolong@uga.edu

Mohammad Nabil El-Nesr, PhD
Researcher, Soil Conservation and Water Resources Dept., Desert Research Center, Egypt; and Associate Professor of Water and Irrigation Systems Engineering, Department of Agricultural Engineering, College of Food and Agricultural Sciences King Saud University, Saudi Arabia. Telephone: +(966) 544909445, KSA. E-mail: drnesr@gmail.com

Mohamed El-Sayed El-Hagarey, PhD
Researcher at Irrigation and Drainage Unit, Division of Water Resources and Desert Land, Desert Research Center (DRC), Ministry of Agriculture and Land Reclamation, 1 Mathaf El-Mataria street, El-Mataria, Cairo-Egypt, B.O.P 11753. Mobile: +2 01063031920. E-mail: elhagarey@gmail.com

Megh R. Goyal, PhD
Retired Professor in Agricultural and Biomedical Engineering from General Engineering Department, University of Puerto Rico – Mayaguez Campus; and Senior Technical Editor-in-Chief in Agriculture Sciences and Biomedical Engineering, Apple Academic Press Inc., PO Box 86, Rincon – PR – 00677 – USA. E-mail: goyalmegh@gmail.com

H. M. Hekal, PhD
Associate Professor of Soil Conservation, Desert Research Center, Egypt

M. M. Hussein, PhD
Researcher Water Relations Dept., National Research Centre (NRC), Cairo, Egypt

Marvin E. Jensen, PhD
P. E., Retired Research Leader at USDA – ARS. 1207 Spring Wood Drive, Fort Collins, Colorado 80525.E-mail: mjensen419@aol.com

A. A. Mady, PhD
Researcher at Water Management and Irrigation Research Institute, Water Research Centre, Cairo, Egypt

Hani A. Mansour, PhD
Researcher, Water Relations Field Irrigation Department, Agricultural and Biological Division, National Research Center, El-Behouth St., Eldokki, Giza, Postal Code 12311. Giza, Egypt. E-mail: hanimansour88@yahoo.com; mansourhani2011@gmail.com

V. M. Mayande, PhD
Vice Chancellor, Dr. Panjabrao Deshmukh Krishi Vidyapeeth, Akola, 444104, Maharashtra-India. Phone: +91 9423174299, E-mail: vmmayande@yahoo.com

Hany M. Mehanna, PhD
Professor, Water Relations and Field Irrigation Dept., National Research Centre, Cairo, Egypt. E-mail: mr.mehana@gmail.com

M. M. Morad, PhD
Professor at Agric. Eng. Dep., Fac. Agric., Zagazig University

Miguel A. Muñoz-Muñoz, PhD
Ex – President of University of Puerto Rico, University of Puerto Rico, Mayaguez Campus, College of Agriculture Sciences, Call Box 9000, Mayaguez, PR. 00681–9000. Tel. 787–265–3871, E-mail: miguel.munoz3@upr.edu

H. Abou-Baker Nesreen, PhD
Researcher at WRFI Dept., Agriculture Division, National Research Centre (NRC), El-Behouth St., Eldokki, Giza, Postal Code 12311, Cairo – Egypt. E-mail: nesreenhaa@yahoo.com

O. Okasha, PhD
Researcher at Water Relations and Field Irrigation Dept., National Research Centre, Cairo, Egypt

M. El-Sayed Omima, PhD
Researcher, Plant Production Dept., Desert Research Center, DRC, Egypt

O. Padmakumari, PhD
Retired Professor in Soil and Water Engineering, College of Agricultural Engineering, Tamil Nadu Agricultural University, Coimbatore – 641003, Tamil Nadu, India

B. J. Pandian, PhD
Dean and Director, Water Technology Center, College of Agricultural Engineering and Technology, Tamil Nadu Agricultural University (TNAU), Coimbatore – 641003, India

B. S. Pathak, PhD
Retired Professor and Dean, College of Agricultural Engineering, Punjab Agricultural University, Ludhiana – India – 141004. E-mail: bspathakprof@gmail.com

Gajendra Singh, PhD
Former Vice – President, Asian Institute of Technology, Thailand. C-86, Millennium Apartments, Plot E-10A, Sector − 61, NOIDA – U.P. – 201301, India, Mobile: (011)-(91) 9971087591, E-mail: prof.gsingh@gmail.com

R. K. Sivanappan, PhD
Former Professor and Dean, College of Agricultural Engineering and Technology, Tamil Nadu Agricultural University (TNAU), Coimbatore. Mailing address: Consultant, 14, Bharathi Park, 4th Cross Road, Coimbatore-641043, India. E-mail: sivanappanrk@hotmail.com

LIST OF ABBREVIATIONS

ASABE	American Society of Agricultural and Biological Engineers
CU	coefficient of uniformity
CV	coefficient of variation
DIS	drip irrigation system
DOY	day of the year
EPAN	pan evaporation
ET	evapotranspiration
ET_c	crop evapotranspiration
FAO	FoodandAgriculturalOrganization, Rome
FC	field capacity
FUE	fertilizers use efficiency
GPIS	gated pipe irrigation system
ICAR	Indian Council of Agriculture Research
ISAE	Indian Society of Agricultural Engineers
LAI	leaf area index
MAD	maximum allowable depletion
MSL	mean sea level
MWD	mean weight diameter
PE	polyethylene
PET	potential evapotranspiration
PM	Penman-Monteith
PVC	poly vinyl chloride
PWP	permanent wilting point
RH	relative humidity
RMAX	maximum relative humidity
RMIN	minimum relative humidity
RMSE	root mean squared error
RS	solar radiation
SAR	sodium absorption rate
SDI	subsurface drip irrigation
SRW	simulated rainwater
SW	Saline water
SWB	soil water balance
TE	transpiration efficiency
TEW	total evaporable water
TMAX	maximum temperature

TMIN	minimum temperature
TR	temperature range
TSS	total soluble solids
TUE	transpiration use efficiency
USDA	US Department of Agriculture
USDA-SCS	US Department of Agriculture-Soil Conservation Service
WSEE	weighed standard error of estimate
WUE	water use efficiency

LIST OF SYMBOLS

A cross sectional flow area (L^2)
AL average life of wells
AW available water (Θ_w, %)
C concentration of chlorine, ppm
c maximum leaf/stem water potential
Cp specific heat capacity of air, in J/(g·°C)
d depth of effective root zone
D depth of irrigation water in mm
E evapotranspiration rate, in g/(m²·s)
e vapor pressure, in kPa
e_a actual vapor pressure (kPa)
Ecp cumulative class A pan evaporation (mm)
eff irrigation system efficiency
E_i irrigation efficiency of drip system
E_p pan evaporation (mm/day)
E_{pan} class A pan evaporation
ER cumulative effective rainfall (mm)
e_s saturation vapor pressure (kPa)
ET evapotranspiration rate, in mm/year
ETc crop-evapotranspiration (mm/day)
EU emission uniformity
F flow rate of the system (GPM)
F.C. field capacity (v/v, %)
G soil heat flux at land surface, in W/m²
gpm gallons per minute
H plant canopy height in meter
h soil water pressure head (L)
I infiltration rate at time t (mm/min)
IR injection rate, GPH
IRR irrigation
K unsaturated hydraulic conductivity (LT^{-1})
K_c cropcoefficient
Kg kilograms
K_p pan coefficient
Kp pan factor
Ks hydraulic conductivity

lph   liters per hour
lps   liters per second
n number of emitters
P   percentage of chlorine in the solution
P.W.P.   permanent wilting point
Pa atmospheric pressure, in Pa
pH   acidity/alkalinity measurement scale
ppm   one part per million
psi   pounds per square inch
Q   flow rate in gallons per minute
q   mean emitter discharges of each lateral (lh^{-1})
R   rainfall
r_a aerodynamic resistance (s m^{-1})
Ra extraterrestrial radiation
R_e effective rainfall depth (mm)
R_i individual rain gauge reading in mm
R_n net radiation at the crop surface (MJ $m^{-2}\,day^{-1}$)
RO   surface runoff
r_s bulk surface resistance (s m^{-1})
S   sink term accounting for root water uptake (T^{-1})
Se effective saturation
S_p plant-to-plant spacing (m)
S_r row-to-row spacing (m)
SU   statistical uniformity (%)
S_ψ water stress integral (MPa day)
t   time that water is on the surface of the soil (min)
T   time in hours
V   volume of water required (liter/day/plant)
V_{id} irrigation volume applied in each irrigation (liter $tree^{-1}$)
V_{pc} plant canopy volume (m^3)
W   canopy width
W_p fractional wetted area
z vertical coordinate positive downwards (L)

GREEK SYMBOLS

α constant lapse rate for moist air
γ psychrometric constant (kPa°C^{-1})
ρa mean air density at constant pressure (kg m^{-3})
Θ_w dry weight basis
Δ slope of the vapor pressure curve (kPa°C^{-1})
θ volumetric soil water content (L^3L^{-3})

$\theta(h)$ soil water retention ($L^3 L^{-3}$),

θr residual water content ($L^3 L^{-3}$)

θ_s saturated water content ($L^3 L^{-3}$)

θ_{vol} volumetric moisture content (cm^3/cm^3)

λ latent heat of vaporization (MJ kg^{-1})

λE latent heat flux, in W/mo

PREFACE

Due to increased agricultural production, irrigated land has increased in the arid and subhumid zones around the world. Agriculture has started to compete for water use with industries, municipalities, and other sectors. This increasing demand along with increments in water and energy costs have made it necessary to develop new technologies for the adequate management of water. The intelligent use of water for crops requires understanding of evapotranspiration processes and use of efficient irrigation methods.

Everyday, news on water scarcity appears throughout the world indicating that government agencies at central/ state/local levels, research and educational institutions, industry, sellers and others are aware of the urgent need to adopt micro irrigation technology that can have an irrigation efficiency up to 90% compared to 30–40% for the conventional irrigation systems. I stress the urgent need to implement micro irrigation systems in water scarcity regions.

Micro irrigation is sustainable and is one of the best management practices. I attended the 17th Punjab Science Congress on February 14–16, 2014 at Punjab Technical University in Jalandhar. I was shocked to learn that the underground water table has lowered to a critical level in Punjab. My father-in-law in Dhuri told me that his family bought the 0.10 acres of land in the city for US $100.00 in 1942 AD because the water table was at 2 feet depth. In 2012, it was sold for US $200,000 because the water table had dropped to greater than 100 feet depth. This has been due to luxury use of water by wheat-paddy farmers. The water crisis is similar in other countries, including Puerto Rico where I live. We can, therefore, conclude that the problem of water scarcity is rampant globally, creating the urgent need for water conservation. The use of micro irrigation systems is expected to result in water savings, and increased crop yields in terms of volume and quality. The other important benefits of using micro irrigation systems include expansion in the area under irrigation, water conservation, optimum use of fertilizers and chemicals through water, and decreased labor costs, among others. The worldwide population is increasing at a rapid rate, and it is imperative that food supply keeps pace with this increasing population.

Micro irrigation, also known as trickle irrigation or drip irrigation or localized irrigation or high frequency or pressurized irrigation, is an irrigation method that saves water and fertilizer by allowing water to drip slowly to the roots of plants, either onto the soil surface or directly onto the root zone, through a network of valves, pipes, tubing, and emitters. It is done through narrow tubes that deliver water directly to the base of the plant. It supplies controlled delivery of water directly to individual plants and can be installed on the soil surface or subsurface. Micro irrigation systems are often used in farms and large gardens but are equally effective in the home garden or even for houseplants or lawns.

The mission of this compendium is to serve as a reference manual for graduate and undergraduate students of agricultural, biological and civil engineering as well as horticulture, soil science, crop science, and agronomy. I hope that it will be a valuable reference for professionals who work with micro irrigation and water management; for professional training institutes, technical agricultural centers, irrigation centers, agricultural extension services, and other agencies that work with micro irrigation programs.

After my first textbook, *Drip/Trickle or Micro Irrigation Management* by Apple Academic Press Inc., and response from international readers, I was motivated to bring out for the world community this ten-volume series on, *Research Advances in Sustainable Micro Irrigation.* This book series will complement other books on micro irrigation that are currently available on the market, and my intention is not to replace any one of these. This book series is unique of its worldwide applicability to irrigation management in agriculture. This series is a must for those interested in irrigation planning and management, namely, researchers, scientists, educators and students.

This book, volume 9, is titled as *Water and Fertigation Managementin Micro Irrigation*, and includes 13 chapters. Water and fertigation management are important practices for the success of drip/micro/trickle irrigation. **Water management** is the activity of planning, developing, distributing, and optimum use of water resources under defined water policies and regulations. It include: management of water treatment of drinking water/industrial water; sewage or wastewater; management of water resources in agriculture; management of flood protection; management of irrigation; management of the water table; and management of drainage, etc. **Irrigation** is the artificial application of water to the land or soil. It is used to assist in the growing of agricultural crops, maintenance of landscapes, and re-vegetation of disturbed soils in dry areas and during periods of inadequate rainfall. Additionally, irrigation also has a few other uses in crop production, which include protecting plants against frost, suppressing weed growth in grain fields, and preventing soil consolidation. In contrast, agriculture that relies only on direct rainfall is referred to as rain-fed or dry-land farming. Irrigation systems are also used for dust suppression, disposal of sewage, and in mining. Irrigation is often studied together with drainage, which is the natural or artificial removal of surface and subsurface water from a given area. Irrigation has been a central feature of agriculture since the start of civilization and is the basis of the economy and society of numerous societies throughout the world. Among all irrigation systems, micro irrigation has the highest irrigation efficiency and is most efficient.

Fertigation is the application of fertilizers, soil amendments, or other water-soluble products through an irrigation system.Chemigation, a related and sometimes interchangeable term, is the application of chemicals through an irrigation system. Chemigation is considered to be a more restrictive and controlled process due to the potential nature of the products being delivered – pesticides, herbicides, and fungicides—to cause harm to humans, animals, and the environment. Fertigation is used extensively in commercial agriculture and horticulture and is starting to be used in general landscape applications as dispenser units become more reliable and easy to use. The irrigator must take into consideration suggestions, such as: (i) Fertigation is used to spoon-feed additional nutrients or correct nutrient deficiencies detected in plant tissue analysis. It is usually practiced on high-value crops such as vegetables, turf, fruit trees, and ornamentals; (ii) Injection during the middle one-third or the middle one-half of the irrigation is recommended for fertigation using micropropagation and drip irrigation; (iii) The water supply for fertigation is to be kept separate from the domestic water supply to avoid contamination; and (iv) The change of fertilizer during the growing season is important in order to adjust for fruit, flower, and root development.

The contribution by all cooperating authors to this book series has been most valuable in the compilation of this volume. Their names are mentioned in each chapter and in the list of contributors of each volume. This book would not have been written without the valuable cooperation of these investigators, many of whom are renowned scientists who have worked in the field of micro irrigation throughout their professional careers.

I would like to thank Sandy Jones Sickels, Vice President, and Ashish Kumar, Publisher and President at Apple Academic Press, Inc., and the editorial staff for making every effort to publish the book when the diminishing water resources is a major issue worldwide. Special thanks are due to the AAP production staff. We request that readers offer us your constructive suggestions that may help to improve the next edition.

I express my deep admiration to my family for understanding and collaboration during the preparation of this ten-volume book series.

With my whole heart and best affection, I dedicate this volume to Dr. Bhim Sen Pathak, who co-founded the College of Agricultural Engineering at Punjab Agricultural Engineering, Ludhiana, India, in 1965; and has been an eminent Professor /Scientist/Dean. Dr. Pathak holds professional and human values I learned from during my college years. He has been my master, councilor, professional father, and guru since 1966. He helped me to trickle on to add my drop to the ocean of service to the world of humanity. Without his advice and patience, I would not have been a *"professional and friendly agricultural and biomedical engineer"* and *"Father of Irrigation Engineering of twentieth century in Puerto Rico,"* with zeal for service to others. My salute to him for his legacy. As an educator, I offer this piece of advice to one and all in the world: *"Permit that our Almighty God, our Creator and excellent Teacher, irrigate the life with His Grace of rain trickle by trickle, because our life must continue trickling on..."*

—Megh R. Goyal, PhD, PE,
Senior Editor-in-Chief
June 10, 2015

FOREWORD 1

With only a small portion of cultivated area under irrigation and with the need to expand this area, which can be brought about by irrigation, it is clear that the most critical input for agriculture today is water. It is important that all available supplies of water should be used intelligently to the best possible advantage. Recent research around the world has shown that the yields per unit quantity of water can be increased if the fields are properly leveled, the water requirements of the crops as well as the characteristics of the soil are known, and the correct methods of irrigation are followed. Significant gains can also be made if the cropping patterns are changed so as to minimize storage during the hot summer months when evaporation losses are high, if seepage losses during conveyance are reduced, and if water is applied at critical times when it is most useful for plant growth.

Irrigation is mentioned in the Holy Bible and in the old documents of Syria, Persia, India, China, Java, and Italy. The importance of irrigation in our times has been defined appropriately by N.D. Gulati: "In many countries irrigation is an old art, as much as the civilization, but for humanity it is a science, the one to survive." The need for additional food for the world's population has spurred rapid development of irrigated land throughout the world. Vitally important in arid regions, irrigation is also an important improvement in many circumstances in humid regions. Unfortunately, often less than half the water applied is used by the crop—irrigation water may be lost through runoff, which may also cause damaging soil erosion, deep percolation beyond that required for leaching to maintain a favorable salt balance. New irrigation systems, design and selection techniques are continually being developed and examined in an effort to obtain high practically attainable efficiency of water application.

The main objective of irrigation is to provide plants with sufficient water to prevent stress that may reduce the yield. The frequency and quantity of water depends upon local climatic conditions, crop and stage of growth, and soil-moisture-plant characteristics. The need for irrigation can be determined in several ways that do not require knowledge of evapotranspiration (ET) rates. One way is to observe crop indicators such as change of color or leaf angle, but this information may appear too late to avoid reduction in the crop yield or quality. Other similar methods of scheduling include determination of the plant water stress, soil moisture status, or soil water potential. Methods of estimating crop water requirements using ET and combined with soil characteristics have the advantage of not only being useful in determining when to irrigate, but also enables us to know the quantity of water needed. ET estimates have not been made for the developing countries though basic information on weather data is available. This has contributed to one of the existing problems that the vegetable crops are over irrigated and tree crops are under irrigated.

Water supply in the world is dwindling because of luxury use of sources; competition for domestic, municipal, and industrial demands; declining water quality; and losses through seepage, runoff, and evaporation. Water rather than land is one of the limiting factors in our goal for self-sufficiency in agriculture. Intelligent use of water will avoid problem of sea water seeping into aquifers. Introduction of new irrigation methods has encouraged marginal farmers to adopt these methods without taking into consideration economic benefits of conventional, overhead, and drip irrigation systems. What is important is "net in the pocket" under limited available resources. Irrigation of crops in tropics requires appropriately tailored working principles for the effective use of all resources peculiar to the local conditions. Irrigation methods include border-, furrow-, subsurface-, sprinkler-, sprinkler, micro, and drip/trickle, and xylem irrigation.

Drip irrigation is an application of water in combination with fertilizers within the vicinity of plant root in predetermined quantities at a specified time interval. The application of water is by means of drippers, which are located at desired spacing on a lateral line. The emitted water moves due to an unsaturated soil. Thus, favorable conditions of soil moisture in the root zone are maintained. This causes an optimum development of the crop. Drip/micro or trickle irrigation is convenient for vineyards, tree orchards, and row crops. The principal limitation is the high initial cost of the system that can be very high for crops with very narrow planting distances. Forage crops may not be irrigated economically with drip irrigation. Drip irrigation is adaptable for almost all soils. In very fine textured soils, the intensity of water application can cause problems of aeration. In heavy soils, the lateral movement of the water is limited, thus more emitters per plant are needed to wet the desired area. With adequate design, use of pressure compensating drippers and pressure regulating valves, drip irrigation can be adapted to almost any topography. In some areas, drip irrigation is used successfully on steep slopes. In subsurface drip irrigation, laterals with drippers are buried at about 45 cm depth, with an objective to avoid the costs of transportation, installation, and dismantling of the system at the end of a crop. When it is located permanently, it does not harm the crop and solve the problem of installation and annual or periodic movement of the laterals. A carefully installed system can last for about 10 years.

The publication of this book series is an indication that things are beginning to change, that we are beginning to realize the importance of water conservation to minimize the hunger. It is hoped that the publisher will produce similar materials in other languages.

In providing this book series on micro irrigation, Megh Raj Goyal, as well as the Apple Academic Press, is rendering an important service to the farmers. Dr. Goyal, *Father of Irrigation Engineering in Puerto Rico*, has done an unselfish job in the presentation of this series that is simple and thorough. I have known Megh Raj since

1973 when we were working together at Haryana Agricultural University on an ICAR research project in "Cotton Mechanization in India."

Dr. Gajendra Singh, PhD, Former Vice Chancellor, Doon University, Dehradun, India. Adjunct Professor, Indian Agricultural Research Institute, New Delhi Ex-President (2010–2012), Indian Society of Agricultural Engineers. Former Deputy Director General (Engineering), Indian Council of Agricultural Research (ICAR), New Delhi. Former Vice-President/Dean/Professor and Chairman, Asian Institute of Technology, Thailand.	 Dr. Gajendra Singh, PhD New Delhi June 10, 2015

FOREWORD 2

Monsoon failure during June of 2014 has created shock waves once again across India. The Indian Meteorological Department has reported a shortage of rains in major parts of India with the country average of 42%, Karnataka 35%, Konkan and Goa 56%, Kerala 24%, Gujarat 88% and Rajasthan 80% during June 2014. India still is 62% agriculture dependent on monsoon rain, and most of the 83% small and marginal farmers are living in these regions. Monsoon failure in June affects food production and livelihood of the majority population of India. The Government of India has taken timely and laudable initiatives to develop a contingency program. India has observed this type of monsoon situation 12 times during the last 113 years, meaning a huge deficit of rain once in 10 years. Although contingency plans provide some relief, there is a need to address fundamental issues of water management in India. India has 1896 km^3 total renewable water resources; in addition only 5% of the total precipitation is harvestable. Improving water productivity is a major challenge. Improving irrigation efficiency, effective rainwater management, and recycling of industrial and sewage water will get enough water available for agriculture in the state. Micro irrigation can mitigate abiotic stress situation by saving over 50% of irrigation water and can be useful in a late monsoon situation for timely sowing.

Agricultural engineers across India have made several specific recommendations on water conservation practices, ground water recharge, improving water productivity, land management practices, tillage/cultivation practices and farm implements for moisture conservation. These technologies have potential to conserve water that will facilitate timely sowing of crops under the delayed monsoon situation that has occurred this year and provide solutions to monsoon worries. Agricultural engineers need to provide leadership opportunities in the water resources and water management sector, which include the departments of Command Area Development, Rural Development, Panchayat Raj, Water Resources, Irrigation, Soil Conservation, Watersheds, Environment and Energy for Stability of Agriculture, and in turn the stable growth of Indian economy.

This book series on micro irrigation addresses the urgent need to adopt this water saving technology not only in India but throughout the world. I would like to see more literature on micro irrigation for use by the irrigation fraternity. I appeal to all irrigation engineering fraternities to bring such issues to the forefront through research publications, organizing symposiums, seminars and discussions with planners and policymakers at the regional, state and national level so that agricultural engineers will get a well-deserved space in the development process of the country.

Dr. V. M. Mayande, PhD
President 2012–15, Indian Society of Agricultural Engineers,
Vice Chancellor,
Dr. Panjabrao Deshmukh Krishi Vidyapeeth
Akola–444104, Maharashtra, India. Tel.: +91 9423174299.
E-mail: vmmayande@yahoo.com

Dr. V. M. Mayande, PhD
June 10, 2015

FOREWORD 3

Irrigation has been a vital resource in farming since the evolution of humans. Due importance to be given for irrigation was not accorded to the fact that the availability has been persistent in the past. Sustained availability of water cannot be possible in the future, and there are several reports across the globe that severe water scarcity might hamper farm production. Hence, in the modern-day farming, the most limiting input being water, much importance is needed for should be given to conservation and the judicious use of the irrigation water for sustaining the productivity of food and other cash crops. Though the availability of information on micro irrigation is adequate, yet its application strategies must be expanded for the larger benefit of the water- saving technology by the users.

In this context under Indian conditions, the attempt made by Prof. R. K. Sivanappan, Former Dean, Agricultural Engineering College of TNAU, in collating all pertinent particulars and assembling them in the form a precious publication proves that the author is continuing his eminent service and support to the farming community by way of empowering them in adopting the micro irrigation technologies at ease and the personnel involved in irrigation also enriched the knowledge of modern irrigation concepts. While seeking the blessings of Dr. R. K. Sivanappan and Dr. Megh Raj Goyal (editor of this book series), I wish the publisher and authors success in all their endeavors, for helping the users of micro irrigation.

—**B. J. Pandian, PhD**

FOREWORD 4

The micro irrigation system, more commonly known as the drip irrigation system, has been one of the greatest advancements in irrigation system technology developed over the past half century. The system delivers water directly to individual vines or to plant rows as needed for transpiration. The system tubing may be attached to vines or placed on or buried below the soil surface.

This book series, written by experienced system designers/scientists, describes various systems that are being used around the world, the principles of microirrigation, chemigation, filtration systems, water movement in soils, soil-wetting patterns, and design principles, use of wastewater, crop water requirements, and crop coefficients for a number of crops. The book series also includes chapters on hydraulic design, emitter discharge and variability, and water and fertigation management of micro irrigated vegetables, fruit trees, vines, and field crops. Irrigation engineers will find this book series to be a valuable reference.

Marvin E. Jensen, PhD, PE
Retired Research Program Leader
at USDA-ARS; and Irrigation Consultant
1207 Spring Wood Drive, Fort Collins,
Colorado 80525, USA
E-mail: mjensen419@aol.com.

Marvin E. Jensen
PhD, PE
June 10, 2015

WARNING/DISCLAIMER

The goal of this compendium, *Water and Fertigation Management in Micro Irrigation*, is to guide the world community on how to manage efficiently for economical crop production. The reader must be aware that dedication, commitment, honesty, and sincerity are the most important factors in a dynamic manner for complete success. This reference is not intended for a one-time reading; we advise you to consult it frequently. To err is human. However, we must do our best. Always, there is a place for learning from new experiences.

The editor, the contributing authors, the publisher, and the printer have made every effort to make this book as complete and as accurate as possible. However, there still may be grammatical errors or mistakes in the content or typography. Therefore, the contents in this book should be considered as a general guide and not a complete solution to address any specific situation in irrigation. For example, one size of irrigation pump does not fit all sizes of agricultural land and will not work for all crops.

The editor, the contributing authors, the publisher, and the printer shall have neither liability nor responsibility to any person, organization, or entity with respect to any loss or damage caused, or alleged to have caused, directly or indirectly, by information or advice contained in this book. Therefore, the purchaser/reader must assume full responsibility for the use of the book or the information therein.

The mention of commercial brands and trade names are only for technical purposes and does not imply endorsement. The editor, contributing authors, educational institutions, and the publisher do not have any preference for a particular product.

All weblinks that are mentioned in this book were active on December 31, 2014. The editors, the contributing authors, the publisher, and the printing company shall have neither liability nor responsibility if any of the weblinks are inactive at the time of reading of this book.

ABOUT THE SENIOR EDITOR-IN-CHIEF

Megh R. Goyal, PhD, PE, is a retired Professor of Agricultural and Biomedical Engineering from the General Engineering Department in the College of Engineering at the University of Puerto Rico – Mayaguez Campus; and Senior Acquisitions Editor and Senior Technical Editor-in-Chief in Agricultural and Biomedical Engineering for Apple Academic Press, Inc.

He received his BSc degree in Engineering in 1971 from Punjab Agricultural University, Ludhiana, India; MSc degree in 1977 and PhD degree in 1979 from the Ohio State University, Columbus; and his Master of Divinity degree in 2001 from Puerto Rico Evangelical Seminary, Hato Rey, Puerto Rico, USA.

Since 1971, he worked as Soil Conservation Inspector (1971); Research Assistant at Haryana Agricultural University (1972–75) and the Ohio State University (1975–1979); Research Agricultural Engineer/Professor at Department of Agricultural Engineering of UPRM (1979–1997); and Professor of Agricultural and Biomedical Engineering at the General Engineering Department of UPRM (1997–2012). He spent a one-year sabbatical leave in 2002–2003 at the Biomedical Engineering Department at Florida International University, Miami, USA.

He was the first agricultural engineer to receive a professional license in Agricultural Engineering in 1986 from the College of Engineers and Surveyors of Puerto Rico. On September 16, 2005, he was proclaimed as "Father of Irrigation Engineering in Puerto Rico for the Twentieth Century" by the American Society of Agricultural and Biological Engineers (ASABE), Puerto Rico Section, for his pioneering work on micro irrigation, evapotranspiration, agroclimatology, and soil and water engineering. During his professional career of 45 years, he has received awards such as: Scientist of the Year, Blue Ribbon Extension Award, Research Paper Award, Nolan Mitchell Young Extension Worker Award, Agricultural Engineer of the Year, Citations by Mayors of Juana Diaz and Ponce, Membership Grand Prize for ASAE Campaign, Felix Castro Rodriguez Academic Excellence, Rashtrya Ratan Award and Bharat Excellence Award and Gold Medal, Domingo Marrero Navarro Prize, Adopted Son of Moca, Irrigation Protagonist of UPRM, and Man of Drip Irrigation by Mayor of Municipalities of Mayaguez/Caguas/Ponce and Senate/Secretary of Agriculture of ELA, Puerto Rico.

He has authored more than 200 journal articles and textbooks on "Elements of Agroclimatology (Spanish) by UNISARC, Colombia," and two bibliographies on "Drip Irrigation." Apple Academic Press, Inc. (AAP) has published many of his books, namely, "Biofluid Dynamics of Human Body," "Biomechanics of Artificial Organs and Prostheses," "Scientific and Technical Terms in Bioengineering and Biotechnology," "Management of Drip/Trickle or Micro Irrigation," "Evapotranspiration: Principles and Applications For Water Management," and "Sustainable Micro Irrigation Design Systems for Agricultural Crops: Practices and Theory." During 2014–2015, AAP is publishing his ten-volume set in *"Research Advances in Sustainable Micro Irrigation."* Readers may contact him at "goyalmegh@gmail.com."

BOOK REVIEWS

"I congratulate the editors on the completion and publication of this book series on micro irrigation. Water for food production is clearly one of the grand challenges of the twenty-first century. "Hopefully this book series will help irrigators and famers around the world to increase the adoption of water savings technology such as micro irrigation. I have known Dr. Goyal since 1982 during our involvement in the US Trickle Irrigation Committee (S143 and W128)."

—**Vincent F. Bralts, PhD, PE**
Professor and Ex-Associate Dean,
Agricultural and Biological Engineering Department,
Purdue University, West Lafayette, Indiana

"This book series is user friendly and is a must for all irrigation planners to minimize the problem of water scarcity worldwide. Dr. Goyal is the *father of irrigation engineering in Puerto Rico of twenty-first century and pioneer on micro irrigation in the Latin America.* Dr. Goyal (my long-time colleague) has done an extraordinary job in the presentation of this series."

—**Miguel A Muñoz, PhD**
Ex-President of University of Puerto Rico;
and Professor/Soil Scientist

"I recall my association with Dr. Megh Raj Goyal while at Punjab Agricultural University, India. I congratulate him on his professional contributions and the distinction in micro irrigation. I believe that this innovative book series will aid the irrigation fraternity throughout the world."

—**A. M. Michael, PhD**
Former Professor/Director,
Water Technology Centre – IARI; Ex-Vice-Chancellor,
Kerala Agricultural University, Thrissur, Kerala

OTHER BOOKS ON MICRO IRRIGATION TECHNOLOGY FROM AAP

Management of Drip/Trickle or Micro Irrigation
Megh R. Goyal, PhD, PE, Senior Editor-in-Chief

Evapotranspiration: Principles and Applications for Water Management
Megh R. Goyal, PhD, PE, and Eric W. Harmsen, Editors

BOOK SERIES: RESEARCH ADVANCES IN SUSTAINABLE MICRO IRRIGATION
Senior Editor-in-Chief: Megh R. Goyal, PhD, PE

Volume 1: Sustainable Micro Irrigation: Principles and Practices
Senior Editor-in-Chief: Megh R. Goyal, PhD, PE

Volume 2: Sustainable Practices in Surface and Subsurface Micro Irrigation
Senior Editor-in-Chief: Megh R. Goyal, PhD, PE

Volume 3: Sustainable Micro Irrigation Management for Trees and Vines
Senior Editor-in-Chief: Megh R. Goyal, PhD, PE

Volume 4: Management, Performance, and Applications of Micro Irrigation
Senior Editor-in-Chief: Megh R. Goyal, PhD, PE

Volume 5: Applications of Furrow and Micro Irrigation in Arid and Semi-Arid Regions
Senior Editor-in-Chief: Megh R. Goyal, PhD, PE

Volume 6: Best Management Practices for Drip Irrigated Crops
Editors: Kamal Gurmit Singh, PhD, Megh R. Goyal, PhD, PE, and
Ramesh P. Rudra, PhD, PE

Volume 7: Closed Circuit Micro Irrigation Design: Theory and Applications
Senior Editor-in-Chief: Megh R. Goyal, PhD; Editor: Hani A. A. Mansour, PhD

Volume 8: Wastewater Management for Irrigation: Principles and Practices
Editor-in-Chief: Megh R. Goyal, PhD, PE; Coeditor: Vinod K. Tripathi, PhD

Volume 9: Water and Fertigation Management in Micro Irrigation
Senior Editor-in-Chief: Megh R. Goyal, PhD, PE

Volume 10: Innovations in Micro Irrigation Technology
Senior Editor-in-Chief: Megh R. Goyal, PhD, PE;
Coeditors: Vishal K. Chavan, MTech, and Vinod K. Tripathi, PhD

BOOK SERIES: INNOVATIONS AND CHALLENGES IN MICRO IRRIGATION
Senior Editor-in-Chief: Megh R. Goyal, PhD, PE

Volume 1: Principles and Management of Clogging in Micro Irrigation
Editors: Megh R. Goyal, PhD, PE, Vishal K. Chavan, and Vinod K. Tripathi

Volume 2: Sustainable Micro Irrigation Design Systems for Agricultural Crops: Methods and Practices
Editors: Megh R. Goyal, PhD, PE, and P. Panigrahi, PhD

Volume 3: Performance Evaluation of Micro Irrigation Management: Principles and practices
Senior Editor-in-Chief: Megh R. Goyal, PhD, PE

HISTORICAL EVOLUTION OF EVAPOTRANSPIRATION METHODS

MARVIN E. JENSEN

CONTENTS

Originally published as Marvin E. Jensen, "Historical evolution of evapotranspiration Methods," Chapter 1, pp. 1–18, In *Evapotranspiration: Principles and Applications for Water Management* by Megh R. Goyal and Eric W. Harmsen (Editors). New Jersey, USA: Apple Academic Press Inc., 2014. Reprinted with permission.

I thank the organizers of Evapotranspiration Workshop on 10th March 2010 by Colorado State University and USDA – ARS for giving me an opportunity to express my personal involvement in the ET research. I ask the readers of this chapter to bear with me, as I have over emphasized my involvement; it is because I have been associated with the development of ET estimating methods for the past 50 years.

1.1   INTRODUCTION

This chapter is a condensation of a more detailed paper that Rick Allen and I prepared in 2000 [53] for the presentation at the Fourth National Irrigation Symposium, ASAE in Phoenix, Arizona. This chapter is also an edited version of my original paper that was prepared for a workshop on Evapotranspiration. This chapter emphasizes my involvement or association with my colleagues in the development and dissemination of new technology for estimating evapotranspiration (ET) in USA. I reviewed current ET literature and older documents relating to the development of early evapotranspiration estimating methods in the USA. In this chapter, I have included some history of the development of the "ASCE Manual 70 Evapotranspiration and Water Requirements [54]," "FAO-56 Crop Evapotranspiration [4],"development of programs to calculate ET using satellite data [5, 9, 10], the new "ASCE Standardized Reference ET Equation [8],"a proposal for a one-step approach to estimate ET, and an update on the second edition of ASCE Manual 70. More detailed but less personnel-oriented, progress in measuring and modeling ET in agriculture can be found in a recent review article by Farahaniet al.[34].

1.2   EARLY STUDIES

With the rapid development of irrigation in the western USA after about 1850, efforts to reduce water losses from both beneficial and nonbeneficial vegetation became more important. Measured water deliveries varied widely and deliveries often greatly exceeded consumptive use. This problem was recognized early on as Buffum[24] stated that over-irrigation was the first and most serious mistake made by early settlers. Most early methods that were developed for estimating evapotranspiration (ET) or consumptive use (CU) for irrigated areas were for seasonal values based on observed or measured water deliveries. Air temperature was the main weather variable that was used. Solar radiation was not considered directly as a separate variable. For monthly values, crop stage of growth effects became important. As developers of empirical estimating methods adjusted and modified their temperature based methods, the methods tended to become more complex. A measured change in soil moisture content over periods of seven or more days was the main source of measured ET data before and during the 1950s. Reliable published data were scarce, especially for measured ET rates by stage of crop growth.

1.3   MY INVOLVEMENT

My involvement in methods of estimating ET began in 1960, when I evaluated monthly coefficients for the Blaney-Criddle equation. In the late 1950s, the United States Department of Agriculture-Soil Conservation Service (USDA-SCS) asked the USDA-Agricultural Research Service (USDA-ARS) to develop monthly coefficients for the Blaney-Criddle(B-C) method. Howard Haise and Harry Blaney had requested USDA-ARS researchers to use their available data to calculate monthly coefficients for the B-C method. A stack of data sheets was collected and given to me when I arrived in Fort Collins in 1959. Before this request, the data had not been analyzed, because Harry Blaney went to Israel for a year and Howard Haise went to University of California, Davis – California for a six-month sabbatical leave. I was asked to work on the data during their absence. I started reviewing the numbers and the results were highly variable. Researchers with little or no prior experience with the B-C method had calculated monthly coefficients. Many did not have the experience to judge good input data from bad data such as drainage that may have occurred following rains. Also, some time periods were too short to be reliable. The plotted monthly coefficients were widely scattered. I revised the questionnaire that was sent to ARS researchers by Howard Haise, as he was well known to ARS researchers at that time. As part of this task, I reviewed current literature and many early publications on development of irrigation and ET estimating methods. These reviews were never published, but served as a good personal reference and the main source of information for a later 1962 ET workshop report [55].

The main difference in the new questionnaire that I prepared was that we asked for basic soil water and weather data, mainly air temperature and precipitation along with supporting data and not the B-C coefficients that they had computed. We established criteria for reviewing datasets before making our own calculations. Where possible, we estimated solar radiation for each measurement period, usually from cloud cover at one or more nearby weather stations. We ended up with about 1000 ET rates for periods of seven or more days. When searching for full crop cover ET data, we found only about 100 datasets represented reasonably reliable values. The general equation [56] summarizing the ET rates and solar radiation (Rs) data from the full cover datasets was developed:

$$ET = [(0.014 \times T) - 0.37] \times (Rs) \qquad (1)$$

where, T is an air temperature in °F, and Rs is solar radiation in mm/day or inches/day. A tabulated summary of the results was presented at an CU workshop that was organized by ARS-SCS in March 1972 and later at the annual meeting of the Soil Sci. Society of America [80]. The workshop report contained: Charts of the ratio of ET to solar radiation for seven crops versus percent of the growing season and between cuttings of alfalfa grown in lysimeters at Reno, Nevada; a summary of solar radiation relationships; and tabulated weekly mean solar radiation, mean air temperature and cloud cover for 20 locations in the western USA. Copies of this workshop report were sent to others involved in estimating ET who did not attend the workshop such as Jerry Christiansen at Utah State University and David Robb at U.S. Bureau of Reclamation (USBR) in Denver – Colorado.

The main results of my research were published in 1963 [56]. The main purpose of this paper was to encourage engineers, soil scientists and agronomists to begin thinking about radiant energy as the primary source of energy for evaporation instead of air temperature, which had been the practice for decades. Using our 1962 report, Dave Robb [75] with the USBR in Denver, Colorado developed a set of nine crop coefficient curves for use with the Jensen–Haise Eq. (1).

1.4   EARLY RESEARCH: IRRIGATION AND CONSUMPTIVE USE

Many books on irrigation had been written on irrigation in England, France and Italy from 1846 to 1888 along with reports on irrigation in California [25]. In 1897, joint efforts were started for irrigation research by USDA – Experiment Stations under the supervision of the Office of Experiment Stations [82]. These investigations were continued for the next 55 years under various USDA departments, but were transferred to the Soil Conservation Service (USDA-SCS) in 1939 [59]. In 1902, the Bureau of Plant Industry was established in the USDA and detailed studies of transpiration were conducted by this organization in the early 1900s.

From about 1890 to 1920, the term duty of water was used to describe the water use in irrigation with no standard definition [48]. Duty of water was sometimes reported as the number of acres that could be irrigated by a constant flow of water such as 1 cubic foot per second or as depth of water applied. Water measured at farm turnouts was referred to as net duty of water. Examples of early studies were those by: Widstoe[93, 94] in Utah from 1902 to 1911 on 14 crops; Harris [41, 42] who summarized 17 years of study in the Cache Valley of Utah; Lewis [61] who conducted studies near Twin Falls – Idaho from 1914 to 1916; Hemphill [44] who summarized studies in the Cache La Poudre river valley of northern Colorado; Israelson and Winsor [47, 48] who made duty of water studies in the Sevier River valley in Utah from 1914 to 1918; and Crandall [28] who worked in the Snake River area near Twin Falls, Idaho in 1917 and 1918. An excellent summary of early seasonal CU studies was presented in a progress report of the Duty of Water Committee in the Irrigation Division of the American Society of Civil Engineers (ASCE). It was presented in 1927 by O.W. Israelson and later published [7].

L.J. Briggs, a biophysicist, and H.L. Shantz, a plant physiologist, conducted highly significant studies of transpiration in eastern Colorado [19–22]. Briggs and Shantz recognized that solar radiation was the primary cause of the cyclic change of environmental factors [23]. They developed hourly transpiration prediction equations using: the vertical component of solar radiation and temperature rise, solar radiation, and vapor saturation deficit. They also recognized the significance of advected energy. A summary of the Briggs and Shantz 1910–1917 studies was later published by Shantz and Piemeizel[76]. Widstoe[92] began studying the influence of various factors affecting evaporation and transpiration in 1902. Harris and Robinson [42] conducted similar studies from 1912 through 1916.Widstoe and McLaughlin [94] concluded that temperature is the most important factor than the sunshine and relative humidity.

Other studies conducted during the 1900–1920 period were related with the factors causing and controlling water loss due to irrigation. During the next two decades emphasis was on the development of procedures to estimate seasonal CU using available climatic data.

1.5   EARLY ESTIMATING METHODS: EVAPOTRANSPIRATION

In 1920, the USBR began studying the relationships between CU and temperature [62]. Hedke[43] proposed a procedure to the ASCE Duty of Water Committee in 1924 for estimating valley CU based on heat available. Heat available was estimated using degree-days. Radiant energy was not considered directly. The ASCE committee concluded in 1927 that there was an urgent need for a relatively simple method of estimating CU [7]. In the 1920s, Harry Blaney began measuring CU of crops based on soil samples. He worked on crops grown along the Pecos River for the Division of Irrigation and Water Conservation under the USDA-SCS. His first procedure for estimating seasonal and annual CU used mean temperature, percent of annual daylight hours and average humidity [17]. From 1937 to 1940, Lowry with the National Resources Planning Board, and Johnson with the USBR, developed a procedure for estimating: Seasonal CU using maximum temperature above 32 °F during the growing season, and annual inflow minus outflow data for irrigation projects [62]. Thornthwaite[84] correlated mean monthly temperature with ET based on eastern river basin water balance studies and developed an equation for potential evapotranspiration which was widely used for years. Thornthwaite recognized the limitations of his equation pointing out the lack of understanding of why potential ET at a given temperature is not the same everywhere. Because of its simplicity, the equation was applied everywhere and, in general, underestimated ET in arid areas. All of these early methods were based on correlations of measured or estimated CU data with various available or calculated climatic data. The resulting equations were relatively simple because computers were not available—only slide rules and perhaps hand-operated adding machines. This may be difficult for young people to comprehend today since they are very dependent on personal computers.

Numerous other reports and publications were prepared between 1920 and 1945 by various state and federal agencies. Many of them dealt with investigations of water requirements for specific areas and measured farm deliveries and not on techniques for estimating CU or ET. Bibliographies of publications on seasonal CU can be found in books such as Israelsen[47] and Houk[45].

1.5.1   BLANEY-CRIDDLE METHOD

In the 1950s and 1960s, the most widely known empirical ET estimating method used in the USA was the Blaney-Criddle(BC) method. The procedure was first proposed by Blaney and Morin in 1942 [17]. It was modified later by Blaney and Criddle[12–15]. The Eq. (2) was well-known older engineers:

$$U = KF = \sum[k \times f] \tag{2}$$

where, U = estimated CU (or ET) in inches; F = the sum of monthly CU factors, f, for the period ($f = t \times p/100$); t = mean monthly air temperature in °F; p = mean monthly percent of annual daytime hours (daytime is defined as the period between sunrise and sunset); K = empirical CU coefficient (irrigation season or growing period); and k = monthly CU coefficient.

For long-time periods mean air temperature was considered to be a good measure of solar radiation [15]. Phelan [73] developed a procedure for adjusting monthly k values as a function of air temperature, which later became part of SCS publication on the BC method [86]. Criddle also developed a table of daily peak ET rates as a function of depth of water to be replaced during irrigation. Hargreaves [38] developed a procedure similar to the BC method for transferring CU data to other areas of the globe. Christiansen and Hargreaves developed a series of regression equations for estimating monthly grass ET based on pan evaporation, air temperature and humidity data [26, 27]. They initiated efforts to reduce weather data requirements to only air temperature, calculated extraterrestrial radiation and the difference between maximum and minimum air temperature to predict the effects of relative humidity and cloudiness. A culmination of these efforts was the well-known 1985 Hargreaves equation for grass reference ET [39, 40].

A summary of early studies and the BC method was published in a USDA technical bulletin in cooperation with the Office of Utah State Engineer [15]. Criddle was Utah State Engineer at that time. This publication presented the BC equation in English and metric units and updated records of measured CU by crops, percent of daytime hours of the year for latitudes of 0 to 65°N. and 0 to 50°S, monthly CU factors by states, suggested monthly crop coefficients (k) for selected locations, monthly CU factors (f) and average precipitation in various foreign countries, and a summary of BC method applications that were made by various consultants such as Claude Fly in Afghanistan, Tipton and Kalmbach in Egypt and West Pakistan. The BC method is still used in some states because historical water records such as water rights have been based on this method.

Prior to about 1960 and U.S. engineers were usually taught only the Blaney-Criddle method of estimating CU or ET. Today, students and young engineers generally have had a fairly broad training in modern methods of estimating ET and ET-climate relationships.

It is unfortunate that Blaney and Criddle did not select extraterrestrial solar radiation as an index of solar energy instead used percent of daytime hours. Daytime hours from sunshine tables of Marvin [65] do not adequately account for effects of solar angle, especially in higher latitudes as illustrated in Fig. 1.1. "Smithsonian Meteorological Tables" would have had an equation for calculating total daily solar radiation at the top of the atmosphere and total daily solar radiation for selected dates during the year at that time.

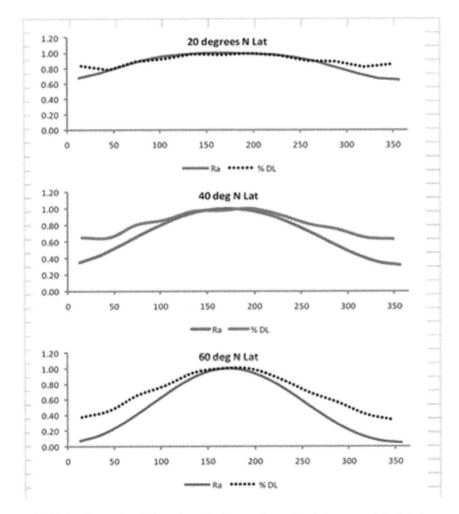

FIGURE 1   Change in relative solar radiation vs. change in relative percent daylight hours.

1.5.2 TRANSITION METHODS IN THE USA

In the 1960 s, estimating ET methods in the USA began to change from methods based primarily on mean air temperature to methods considering both temperature and solar radiation. Several of these methods are listed below:

Alfalfa reference ET	Jensen and Haise[56]	$ET = ((0.014 \times T - 0.37)) \times (R_s)$ (3)
Grass reference ET	Hargreaves and Samani[40] and Hargreaves et al.[39]	$ET = (0.0023) \times R_a \times (T + 17.8) \times (TD)^{0.50}$ (4)
Grass reference ET, Florida	Stephens and Stewart (1963)	$ET = (0.0082 \times T_f - 0.19) \times (R_s)$ (5)
where, ET=Potential evapotranspiration inmm/day, T is a mean air temperature in °C, R_s is solar radiation in mm per day or inches per day, R_a =Extraterrestrial radiation, mm/ day, and TD is the difference between maximum and minimum daily air temperature in °C.		

1.5.3 THEORETICAL METHODS

The Bowen ratio is the ratio of temperature to vapor pressure gradients, $\Delta t / \Delta e$, [18, 26]. Bowen ratio and energy balance concepts were not incorporated at an early date into methods for estimating ET as they were for estimating evaporation from water surfaces. Examples of early work on evaporation from water using the Bowen ratio were those of Cummings and Richardson [31], McEwen [96], Richardson [97], Cummings [29], Kennedy and Kennedy [58] and Cummings [30].

   In contrast to development of largely empirical methods in the USA, Penman [67] in the United Kingdom took a basic approach and related ET to energy balance and rates of sensible heat and water vapor transfer. Penman's work was based on the physics of the processes, and it laid the foundation for current ET estimating methodology using standard weather measurements of solar radiation, air temperature, humidity, and wind speed. The Penman equation [69–71] stands out as the most commonly applied physics-based equation. Penman [69] used the Bowen ratio principle in developing the well-known equation for estimating evaporation from a water surface along with reduction coefficients for grass. Later, a surface resistance term was added [66, 72, 74]. Howell and Evett[46] provided an excellent summary of the Penman equation and its evolution to the development of a standardized equation by the American Society of Civil Engineers. The modern combination equation applied to standardized surfaces is currently referred to as the Penman-Monteithequation (PM). It represents the state-of-the-art in estimating hourly and daily ET. When applied to standardized surfaces, it is now called the Standardized Reference ET Equation [8].

Other methods of estimating and measuring ET range from eddy covariance and energy balance using Bowen ratio or sensible heat flux based on surface temperature, radiosonde measurements of complete boundary layer profiles of temperature, humidity and energy balance estimates based on satellite imagery.

A **radiosonde** (*Sonde* is French and German for *probe*) is a unit for use in things such as weather balloons that measures various atmospheric parameters and transmits them to a fixed receiver. Radiosondesmay operate at a radio frequency of 403 MHz or 1680 MHz and both types may be adjusted slightly higher or lower as required. A **rawinsonde** is a radiosonde that is designed to only measure wind speed and direction. Colloquially, rawinsondes are usually referred to as radiosondes. Modern radiosondes measure or calculate the following variables: Pressure, Altitude, Geographical position (Latitude/ Longitude), Temperature, Relative humidity, Wind (both wind speed and wind direction), Cosmic ray readings at high altitude. Radiosondes(http://en.wikipedia.org/wiki/Radiosonde) measuring ozone concentration are known as ozonesondes.

1.5.4 OTHER METHODS

During the 1950s and early 1960s, many other equations were proposed, but they have not been widely used in the USA. Some of these are Halkiaset al.[37], Ledbedevich[60], Romanov [98], Makkink[63] and Vitkevich[90]. In 1957, Makkink published a formula for estimating potential ET based on solar radiation and air temperature [74] that is still used in Western Europe. Makkink used the energy-weighting term of the Penman equation, solar radiation and a small negative constant. The Makkink equation formed the basis of the subsequent FAO Radiation method that was included in FAO Irrigation and Drainage Paper No. 24 on crop water requirements by Doorenbos and Pruitt [33, 34]. Turc developed a formula in 1960, which was later modified [85]. It was based on mean air temperature and solar radiation for 10-day periods. Rijtema[74] proposed the Turc formula for individual crops using crop factors and length of growing season. Olivier [68] in England developed a procedure for estimating average monthly CU for planning new projects where climatic data were limited. The Olivier's equation used average monthly wet-bulb depression and a factor based on clear sky solar radiation values.

In 1968, I described the process of using the rate of ET from a well-watered crop with an aerodynamically rough surface like alfalfa with 30–50 cm of growth as a measure of potential ET or E_o[50]. ET for a given crop could be related to E_o using a coefficient, now commonly known as a crop coefficient:

$$ET = [K_c] \times [E_o] \tag{6}$$

where, K_c is a dimensionless crop coefficient similar to that proposed by van Wijk and de Vries[89] representing the combined effects of resistance to water movement from the soil to the evaporating surfaces, resistance to diffusion of water vapor from the evaporating surfaces through the laminar boundary layer, resistance to turbulent transfer to the free atmosphere, andrelative amount of radiant energy available as compared to the reference crop. At that time, methods other than those based on air temperature were not well known. In order to facilitate the understanding of the of the "($E_o \times K_c$) process,"illustrating the change in the K_c as crop cover develops enabled users to visualize how the coefficient changed from a value near 0.15 for bare soil to 1.0 at full cover.

Routine use of computers for estimating ET was in its infancy. Estimates of alfalfa reference ET and ETr, that were used in our first computerized irrigation scheduling program [51] were calculated using the Penman method with alfalfa wind speed coefficients developed at Kimberly, ID [57]. Since we did not have a computer at Kimberly – Idaho, we used a time-share computer located in Phoenix – Arizona connected via the telephone system with a teletype paper tape reader and printer at Kimberly. In 1970, copies of a FORTRAN computer program that Ben Pratt and I had written for estimating daily ET and scheduling irrigations using the Penman equation was widely distributed following an informal workshop at Kimberly – Idaho in 1970.

The method of using reference ET and crop coefficients has been widely used for nearly a half century. In general, it has been relatively robust. For example, in 2004, Ivan Walter and I estimated ET for the Imperial Irrigation District in California using this approach. Our estimate of ET for agricultural land for CY 1998 was 2% higher than that of a SEBAL estimate for entire district; and 5% higher for agricultural crops for water year 1998 [83] of 2.01 million ac-ft. or 2.5 km^3. Whether the [$ET_{ref} \times K_c$] method, also known as the two-step method, will be replaced by the direct PM method, or one-step method, in the near future is uncertain.

Various methods of measuring ET and methods of estimating ET were summarized during a conference held in Chicago following the ASAE winter meeting on Dec. 5–6, 1966. Leading researchers from many different organizations and disciplines presented papers on the current state-of-the-art of estimating ET in the early 1960s. I was the conference chairman and program committee members represented ASCE, American Meteorological Society, Soil Sci. Society of America, International Commission of Irrigation and Drainage, and several Canadian organizations. W.O. Pruitt assisted in editing the proceedings [49].

Unfortunately, we not did get anyone to specifically discuss the Penman method. I tried to get Howard Penman to attend and present a paper, but he declined after several letters and a phone call. Leo Fritschen discussed the energy balance method, C.H.M. van Bavel discussed combination methods [87, 88], and Champ Tanner discussed a comparison of energy balance and mass transfer methods for measuring ET [81]. The conference proceedings contained pictures of the authors who were leaders in their field at the time and session chairmen.

1.6   DISSEMINATING AND ADOPTING NEW ESTIMATING TECHNOLOGY FOR EVAPOTRANSPIRATION

Disseminating and adopting advances in ET estimating technology has not been rapid as compared with many other fields because dissemination-adaption involves many disciplines such as agricultural engineering, hydrology, water science, meteorology, soils, and plants. Dissemination and adaption of new technology did not occur rapidly as it did in single disciplines. The users of the technology, a half-century ago, were mainly engineers.

Penman had a comprehensive understanding of the physical processes involved in ET. He presented an introductory paper at the informal meeting on physics that was held in The Netherlands in September 1955 [70]. He concluded with the sentence: *"Though the physicist still has some problems he can solve by himself, much of his future contribution to understanding evaporation in agriculture must be in collaboration with the biologist and the soil scientist."*

In my 1960 review of old and current literature, it was clear that new advances in estimating ET had to depart from the traditional use of temperature as the primary input variable. The Penman equation had been developed, but was not in general use by engineers designing and managing irrigation projects. Generally, the Penman equation was thought to be too complicated for use by engineers given the status of computational tools and weather data commonly collected at that time. In developing countries, it is still considered complex.

During the 1966 meeting of the Irrigation and Drainage Division of ASCE in Las Vegas, Nevada, I was asked to chair the ASCE committee on Irrigation Water Requirements, formerly the Committee on CU of Crops and Native Vegetation. The committee was charged with developing a manual on CU. The committee had been chaired by Harry Blaney, but it was not making much progress on this task. A few papers that had been prepared were reproductions of old temperature based methods of estimating ET. My one condition on accepting chairmanship was that I could bring in non-ASCE members into the committee such as Bill Pruitt. Bill, who was with the University of California at Davis and was not an ASCE member then. Bill had been measuring ET using weighing lysimeters and collecting associated basic meteorological data. The committee had control of the Corresponding and Non-member advisors. Control members were: Robert Burman, University of Wyoming; Harlan Collins, SCS; Albert Gibbs, USBR; Marvin E. Jensen, ARS-USDA; and Arnold Johnson, USGS. Harry Blaney remained on the committee, but never attended any of our meetings. We made progress and produced a report that was the start of the ASCE manual on ET. In 1973, we prepared as a camera-ready document at Kimberly – Idaho that was printed as the ASCE report Consumptive Use and Irrigation Requirements [52] and widely disseminated in the U.S. and in other countries like China.

I left the ET committee for three years when I became a member of the executive committee of the ASCE-I&D Division in October 1974. Members of the ET committee continued to work on the manual. Several members served as ET committee chairman between October 1973 and 1986. In 1986, a subcommittee of the ET committee was formed. Its members were Rick Allen, Ron Blatchley, Bob Burman, Marvin E. Jensen, Eldon Johns, Jack Stone and Jim Wright. I was designated as chairman. We were charged with the task of preparing the CU or ET manual. The first draft, which had both English and metric units, was completed for review by one or several ASCE committees in 1988. One reviewer suggested that we use only metric units. This was a good advice, but delayed the manuscript another year as changes were made in the manuscript. The manual was published in 1990 as ASCE Manual No. 70 [54]. The ASCE-PM equation from ASCE Manual 70 has been widely accepted for standardized calculations such as in the Natural Resources Conservation Service National Engineering Handbook [64].

Allen et al.[2] prepared a paper for publication in the Agron. J. to disseminate new information to soil scientists and agronomists. In 1990, the FAO needed to update its 1975–1977 publication on crop water requirements [4]. It organized an expert's consultation on the revision of FAO-24 in Rome on 28–30 May 1990 [79]. There were 10 participants from seven countries: John Monteith from the International Crops Research Institute for Semi-arid Tropics, D. Rijks from WMO, and several participants from FAO. USA participants were Rick Allen, Marvin E. Jensen, and Bill Pruitt. At this conference several manuscript copies of the ASCE Manual 90 were made available and became a key reference. The revision of FAO-24 with major contributions by Rick Allen resulted in the well-known FAO-56 publication on crop evapotranspiration by Allen et al.[4].

In the Netherlands, R.A. Feddes had been working on models of ET, plant root systems and soil water extraction [35]. In 1998, Bastiaanssen et al., including Feddes, published a paper describing a surface energy balance model for land (SEBAL) using satellite- based imagery [9, 10]. In the USA, Allen et al., developed a high resolution-mapping model with internalized calibration using principles and similar techniques as used in the SEBAL model [5]. It uses a near-surface temperature gradient (dT) that is indexed to the radiometric surface as in SEBAL. The METRIC technique [5] uses the SEBAL technique for estimating dT. METRIC also uses weather-based reference ET to establish energy balance conditions for a cold pixel and is internally calibrated at two extreme conditions (wet and dry) using local available weather data. The auto-calibration is done for each image using alfalfa-based reference ET. Both of these models are now being used in the U.S. to map ET at a high resolution over large areas.

In 2000, the Irrigation Association and landscape industry requested the ASCE Irrigation Water Requirements committee, now renamed Committee on Evapotranspiration in Irrigation and Hydrology, to recommend a single procedure for estimating reference ET for use in the USA. This request in part resulted in an ASCE task committee of the ET Committee consisting of engineers and scientists from around the USA. They agreed on a single equation for estimating reference crop ET. The equation is a simplification of the ASCE-PM equation. The request was made to help standardize the basis for the myriad of landscape (i.e., crop) coefficients that have been developed since the late 1980 s. The task committees suggested applying the PM equation to both a tall reference crop like alfalfa and short reference crop like clipped grass by changing several coefficients in order to support usage in both agricultural and landscape industries. A reduced form of the PM equation was adopted for both reference types, with the grass form being the same as in the FAO-56 publication. This was done to promote agreement in usage between the USA and other countries. Sets of coefficients were presented in a table for estimating daily or hourly reference ET for the short and tall references. Details of that equation and its development were presented in separate papers at the 2000 ASAE National Irrigation Conference in Phoenix, Arizona [91]. The various forms and applications of the PM and Penman equations, as well as commonly used empirical equations, were implemented in REF-ET softwarethat was available for free downloading from: http://www.kimberly.uidaho.edu/ref-et, Current Version: 3.1.08, updated January, 2012.

Rick Snyder also has several reference ET programs (monthly, daily, and hourly) for Excel spreadsheets on the web site: http://biomet.ucdavis.edu/evapotranspiration.html.

A summary of the reference ET methodology and tables of mean and basal crop coefficients were published as a major chapter in the 2nd edition of the ASABE book Design and Operation of Farm Irrigation Systems [6]. It contains a great deal of detail on factors controlling ET and on estimating ET. There are numerous recent publications on many different crop coefficients written by authors in the USA and other countries. Before using these coefficients, the users need to carefully review how the calculations were made. FAO-56 crop coefficients have been refined incorporating the fraction of ground cover and plant height [3]. Others have used remote sensing with ground-based or aircraft-based cameras, and satellite images to measure the normalized difference vegetation index (NDVI) and then related NDVI to ET_{ref}-based crop coefficients. The use of the reference ET and crop coefficient method is expected to continue because of the extensive collection of available crop coefficients.

Some scientists are suggesting a more direct approach to estimating crop water requirements such a one-step method or direct PM [66]. Shuttleworth derived a Penman-Monteithbased, one-step estimation equation that allows for different aerodynamic characteristics of crops in all conditions of atmospheric aridity to estimate crop ET from any crop of a specified height using standard 2-m climate data [77]. Not everyone agrees that the concept can adequately account for surface soil water conditions. Shuttleworth called for field studies to address the problem of effective values for surface resistance for different crops equivalent to that for crop coefficients. Shuttleworth and Wallace summarized a detailed study that was conducted in Australia using the one-step approach [78]. I am not aware of any specific applications that have been made in the USA. In the second edition of Manual 70, we have a chapter on "Direct Penman-Monteith method."

The first edition of Manual 70 is out of print. In 2000, the ET committee decided to have a technical committee prepare a second edition. Marvin E. Jensen and Rick Allen were designated cochairmen. Other members were Terry Howell, Derrel Martin, Rick Snyder, and Ivan Walter. The second edition has been completely restructured. We are near having a final draft ready for review by ASCE committees. I hope that this can be completed this year.

Major progress on developing improved methods of estimating ET was made during the past third of a century because of the efforts of many Europe and U.S. individuals. Some are R. Feddes and W. Bastiaannsen in the Netherlands; L. Pereira in Portugal; and M. Smith with FAO who led the FAO effort. Many individuals in the U.S. were major contributors to ET development technology. Some of these are Rick Allen, Terry Howell, Bill Pruitt, Joe Ritchie, Rick Snyder and Jim Wright. Bill Pruitt measured ET in weighing lysimeters near Davis – California along with detailed weather data for many years. His technical guidance and data were very valuable in the development of new technology. Jim Wright measured ET from various crops using a weighing lysimeter at Kimberly, ID over an eight-year period and he developed the concept of the basal crop coefficient representing conditions when soil evaporation is minimal and most of the ET is transpiration [95]. Wright's measured data were also used to refine net radiation and crop coefficients.

1.7   SUMMARY

This chapter summarizes a century of progress in the development of modern methodology for accurately estimating daily and hourly evapotranspiration. Why did this process take so long? Evaporation from soil and plant services is a complex process involving plants, soils, local weather data like wind speed and humidity, solar and long-wave radiation. Its developments involved many disciplines. It has only been 200 years since hydrologic principles were first understood and described by Dalton so perhaps a century is not that unreasonable.

Most of the progress in developing new methodology in the U.S. was made during the last third of a century. Many scientists and engineers were involved in the evolution of ET-estimating technology. Scientists and engineers in Europe have also been instrumental in advancing the technology. In this chapter, I tried to highlight major contributions of people many of whom I knew personally or at least I had met them briefly. My involvement in the process of development better ET estimating technology and association with other engineers and scientists was a learning experience. It started me on a very rewarding career path. The experience that I gained working leading scientists and engineers over the past half century has been very rewarding personally. If I have emphasized my involvement too much it is because I have been associated with the development of ET estimating methods for the past

50 years. More detailed progress on measuring and modeling ET can be found in a review paper by Farahaniet al.[34].

KEYWORDS

- **Blaney-Criddle**
- **Bowen ratio**
- **consumptive use**
- **crop coefficient**
- **crop water requirement**
- **duty of water**
- **energy balance**
- **evaporation**
- **evapotranspiration**
- **FAO**
- **ground cover**
- **irrigation**
- **irrigation engineering**
- **meteorology**
- **Penman – Monteith**
- **satellite-based imagery**
- **SEBAL**
- **soil moisture**
- **solar angle**
- **solar radiation**
- **transpiration**
- **USDA-ARS**
- **USDA-SCS**
- **water requirement**

REFERENCES

1. Allen, Richard G.(2009). Estimating crop coefficients from fraction of ground cover and height. *Irrig. Sci.28,* 17–34.
2. Allen, R. G., Jensen, M. E., Wright, J. L.,Burman, R. D., Operational estimates of evapotranspiration. *Agron. J.* (1989). *81(4),* 650–662.
3. Allen, R. G., Pereira, L. S.(2009). Estimating crop coefficients from fraction of ground cover and height. *Irrig. Sci.28,* 17–34.

4. Allen, R. G., Pereira, L. S.,Raes, D., Smith,M. (1998). Crop Evapotranspiration: Guidelines for Computing Crop Water Requirements. United Nations Food and Agriculture Organization, Irrigation and Drainage Paper No. 56, Rome, Italy. 300 pages.
5. Allen, R. G.,Tassumi, M.,Trezzo,R. (2007). Satellite-based energy balance for mapping evapotranspiration with internalized calibration (METRIC)—Model. J. Irrig. and Drain. Engr. *133(4)*, 380–406.
6. Allen, R. G., Wright, J. L., Pruitt, W. O., Pereira, L. S.,Jensen, M. E. (2007). Water requirements. Chap.8, pp. 208–288, In: *Design and Operation of Farm Irrigation Systems*. Second ed.; 875. Am. Soc. Agri. Biol. Engrs, St. Joseph, MI.
7. Anonymous, (1930).Consumptive Use of Water in Irrigation.Prog. Rep.; Duty of Water Comm.; Irrig. Div.; Trans. of ASCE, *94*, 1349–1399.
8. ASCE-EWRI, (2005). *The ASCE Standardized Reference Evapotranspiration Equation*.Edited by Allen, R. G., Walter, I. A., Elliott, R. L., Howell, T. A.,D.,Itenfisu, Jensen, M. E.,Snyder, R. L., Am. Soc. Civ. Engrs.; 69 pages with appendices A to F Index.
9. Bastiaanssen, W. G. M.,Mensenti, M.,Feddes, R. A.,Holtslag, A. A. M. (1998).A remote sensing surface energy balance algorithm for land (SEBAL), part1, Formulation. J. Hydrology, *212*, 198–212
10. Bastiaanssen, W. G. M.,Pelgrum, H., Wang, J., Ma, Y., Moreno, J.,Roerink, G. J.(1998). Surface energy balance algorithm for land (SEBAL), part2, Validation. J. Hydrology, *212*, 213–229.
11. Blaney, H. F. (1952).Definitions, methods, and research data: consumptive use of water.A symposium, Trans. Am. Soc. Civ. Engr. *117*, 849–973.
12. Blaney, H. F.,Criddle, W. D. (1945). Determining water requirements in irrigated areas from climatological data. Processed, 17 p.
13. Blaney, H. F.,Criddle, W. D. (1945). A method of estimating water requirements in irrigated areas from climatological data. (mimeo).
14. Blaney, H. F.,Criddle, W. D. (1950). Determining water requirements in irrigated areas from climatological and irrigation data. USDA-SCS-TP-96 Report, 50 pages.
15. Blaney, H. F.,Criddle, W. D. (1962). Determining Consumptive Use and Water Requirements. USDA, Technical Bull. No. 1275. 63 pages.
16. Blaney, H. F., Rich, L. R.,Criddle, W. D. (1952).Consumptive use of water. Trans. Am. Soc. Civ. Engr. *117*, 948–1023.
17. Blaney, H. F.,Morin, K. V. (1942).Evaporation and consumptive use of water formulas. Am. Geophys. Union Trans. *1*, 76–82.
18. Bowen, I. S.,The ratio of heat losses by conduction and by evaporation from any water surface. *Physics Review*, (1926). *27*, 779–787.
19. Briggs, L. J. Shantz, H. L. (1913).The water requirements of plants: I., Investigations in the Great Plains in 1910 and 1911. USDA *Bur. Plant Indr. Bull.284*, 49 pp.
20. Briggs, L. J. Shantz, H. L.,Relative water requirements of plants. *J. Agr. Res. (1914).III(1)*, 11–64.
21. Briggs, L. J. Shantz, H. L. (1916a). Hourly transpiration rate on clear days as determined by cyclic environmental factors. J. Agr. Res.; *5(14)*, 583–648.
22. Briggs, L. J. Shantz, H. L. (1916b). Daily transpiration during the normal growth period and its correlation with the weather. J. Agr. Res.; *7(4)*, 155–212.
23. Briggs, L. J. Shantz, H. L. (1917).A comparison of the hourly transpiration rate of atmometers and free water surfaces with the transpiration rate of Medicago sativa. J. Agr. Res. *9(9)*, 279–292.
24. Buffum, B. C. (1892). Irrigation and duty of water. Wyo. Agr. Exp. Bull. No. 8.
25. Carpenter, L. G. (1890). Section of meteorology and irrigation engineering. 3rd Ann. Report of Colorado Exp. Stn.; Fort Collins.

26. Christiansen, J. E.Pan evaporation and evapotranspiration from climatic data. J. Irrig. Drain. Div. (1968). *94*, 243–256.

27. Christiansen, J. E., Hargreaves, G. H. (1969).Irrigation requirements from evaporation. Trans. Int. Comm. Irrig. Drain. *3(23)*, 569–23.596.

28. Crandall, L. (1918). Report of use of water on Twin Falls North Side (unpublished data).

29. Cummings, N. W. (1936). Evaporation from water surfaces. Am. Geophys. Union Trans.; Part2, 507–509.

30. Cummings, N. W. (1940).The evaporation-energy equations and their practical application. Am. Geophys. Union Trans. *21*, 512–522.

31. Cummings, N. W., Richardson,B. (1927). Evaporation from lakes. Physics. Rev. *30(4)*, 527–534.

32. Doorenbos, J. Pruitt, W. O. (1975).Guidelines for predicting crop water requirements. FAO Irrig. Drain. Paper No.24,FAO, Rome, 179 pp.

33. Doorenbos, J. Pruitt, W. O. (1977).Guidelines for predicting crop water requirements. FAO Irrig. Drain. Paper No. 24 (Revised), FAO, Rome, 179 pp.

34. Farahani, H. J. Howell, T. A.,Shuttleworth W. J., Bausch, W. C. (2007).Evapotranspiration: Progress in measurement and modeling in agriculture. Trans. of the ASBE, *50(5)*, 1627–1638.

35. Feddes, R. A.,Menenti, M.,KabatP.,Bastiaanssen, W. G. M. (1993).Is large scale modeling of unsaturated flow with areal average evaporation and surface soil moisture as estimated from remote sensing feasible? J. Hydrol. *143*, 125–152.

36. Fritschen, L. J. (1966). Energy Balance Method. 34–37, In: Jensen, M. E. (ed), *Proceedings on Evapotranspiration and its Role in Water resources Management.* Am. Soc. Agr. Engr.; Chicago, IL, December 5–6, 68 pp.

37. Halkias, N. A.,Veihmeyer, F. J.,Hendrickson, A. H. (1955).Determining water needs for crops from climatic data. Hilgardia, *24(9).*

38. Hargreaves, G. H. (1956).Irrigation requirements based on climatic data. Proc. Am. Soc. Civ. Engr.; Irrig. Drain. Div.; 82(IR3), Paper 1105, 1–10.

39. Hargreaves, G. L., Hargreaves, G. H., Riley, J. P. (1985).Agricultural benefits for Senegal river basin. J. Irrig. Drain. Engr.; Am. Soc. Civ. Engr. *111(2)*, 113–124.

40. Hargreaves, G. H.,Samani, Z. A. (1985).Reference crop evapotranspiration from temperature. Applied Eng. in Agr. Am. Soc. Agr. Engr, *1(2)*, 96–99.

41. Harris, F. S. (1920). The duty of water in the Cache Valley Utah. Utah Agr. Exp. Stn. Bull. Number 173.

42. Harris, F. S., Robinson, J. S. (1916).Factors affecting the evaporation of moisture from soil. J. Agr. Res. *7(10).*

43. Hedke, C. R. (1924). Consumptive use of water by crops. New Mexico State Engr. Office, July.

44. Hemphill, R. G. (1922). Irrigation in Northern Colorado. USDA Bull. Number 1026.

45. Houk, I. E. (1951). Irrigation Engineering,1, Agricultural and Hydrological Phases. John Wiley & Sons, Inc.; New York.

46. Howell, T. A.,Evett, S. R. (2002). The Penman-Monteith method.www.cprl.ars.usda.gov.

47. Israelsen, O. W. (1950). Irrigation Principles and Practices. 2nd ed.; John Wiley & Sons, Inc.; New York, 405.

48. Israelsen, O. W., Winsor, L. M. (1922). The net duty of water in the Sevier Valley. Utah. Utah Agr Exp. Stn. Bull. 182.

49. Jensen, M. E. (ed),(1966). Proceedings on Evapotranspiration and its Role in Water resources Management. Am. Soc. Agr. Engr.; Chicago, Dec. 5–6, 68 pp.

50. Jensen, M. E.Water consumption by agricultural plants. In: T. T., Kozlowski (ed.), *Water Deficits and Plant Growth*: Academic Press, New York. (1968). *2*, 1–22.

51. Jensen, M. E.Scheduling irrigations with computers. J. Soil and Water Conservation, (1969). *24(5)*, 193–195.

52. Jensen, M. E. (ed),(1974). Consumptive use of Water and Irrigation Water Requirements. Rept. Tech. Com. on Irrig. Water Req.; Irrig. Drain. Div.; Am. Soc. Civ. Engr, 229 pp.

53. Jensen, M. E.,Allen, R. G. (2000).Evolution of practical ET estimating methods.In Evans, R. R., B. L.,Benham and T. P.,Trooien(eds), National Irrigation Symposium, Am. Soc. Agr. Engr.; St. Joseph, MI. 52–65.

54. Jensen, M. E.,Burman R. D., Allen R. G. (eds),Evapotranspiration and Irrigation Water Requirements, Am. Soc. Civ. Engr. Manuals and Repts. Eng. Practice No.70,ISBN 0-87262-763-2, (1990). 360 pp.

55. Jensen, M. E.,H. R Haise, (1962). Estimating Evapotranspiration from Solar Radiation. Prel. Rept. for discussion at an ARS-SCS Consumptive Use Workshop, Phoenix, AZ, Mar. 6–8.

56. Jensen, M. E.,Haise, H. R. (1963). Estimating evapotranspiration from solar radiation. Proc. J. Irrig. Drain. Div.; Am. Soc. Civ. Engr.; 89(IR4), 15–41, Closure, 91(IR1), 203–205.

57. Jensen, M. E., D. Robb, C. N.,Franzoy, C. E. (1970).Scheduling irrigations using climate-crop-soil data. J. Irrig. Drain. Div.; Am. Soc. Civ. Engr. *96*, 25–28.

58. Kennedy, R. E.,Kennedy, R. W. (1936).Evaporation computed by the energy-equation. Am. Geophys. Union Trans. *17*, 426–430.

59. Knoblauch, H. C., Law, E. N., Mayer, W. P. (1962). State Experiment Stations. USDA Misc. Publ. 904.

60. Ledbedevich, N. F. (1956). Water regime in peat and swamp soils in the Belo Russian and crop yields. Belo Russian, S. S. R., 6 (translated from Russian), USDC PST Cat.489,1961.

61. Lewis, M. R. (1919). Experiments on the proper time and amount of irrigation. Twin Falls Exp. Stn. 1914, 1915, and 1916.

62. Lowry, R. L., Jr. and Johnson, A. F. (1942). Consumptive use of water for agriculture. Trans. Am. Soc. Civ. Engr.; *107*, 1243–1302.

63. Makkink, G. F.(1957). Testing the Penman formula by means of lysimeters. J. Inst. Water Eng. *11(3)*, 277–278.

64. Martin, D. L.,J. Gilley, (1993). Irrigation Water Requirements.Chapter *2*, Part 623. In: *National Engineering Handbook*. USDA – Soil Conservation Service. 284 pages.

65. Marvin, C. F. (1905). Sunshine Tables. U. S., Weather Bur. Bull. *805*, (Reprinted 1944). McEwen, G. F. (1930). "Results of evaporation studies conducted at the Scripps Institute of Oceanography and the California Institute of Technology." Bull. the Scripps Inst. of Oceanography, Techn. Series, *2(11)*, 401–415.

66. Monteith, J. L. (1965). Evaporation and the environment. p. 205–234, In: *The State and Movement of Water in Living Organisms*. XIXth Symposium. Soc. for Exp. Biol.; Swansea, Cambridge Univ. Press.

67. Monteith, J. L. (1986). Howard Latimer Penman, 10 April (1909). to 13 October (1984). Biographical Memoirs of Fellows of the Royal Society, 32 (Dec. 1986), 379–404. Stable URL: http: //www.jstor.org/stable/770117.

68. Olivier, H. (1961). Irrigation and Climate. Edward Arnold Publ.; LTD., London.

69. Penman, H. L.(1948). Natural evaporation from open water, bare soil and grass. Proc. Roy. Soc. London, Series, A. *193*, 120–146

70. Penman, H. L. (1956).Evaporation: An introductory survey. Neth. J. Agr. Res. *4(1)*, 9–29.

71. Penman, H. L. (1963). Vegetation and Hydrology. Tech. Common.*53*, Commonwealth Bureau of Soils, Harpenden, England.

72. Penman, H. L., Long, I. F. (1960).Weather in wheat: An essay in micrometeorology. Qtrly. J. Roy. Meteorol. Soc. *86*, 16–50.

73. Phelan, J. T. (1962). Estimating monthly "k" values for the Blaney-Criddle formula. ARS-SCS Workshop on Consumptive Use, Mar. 6–8(mimeo).

74. Rijtema, P. E. (1965). Analysis of actual evapotranspiration.Agr. Res. Rep. No. 69. Centre for Agr.Publ. and Doc.; Wageningen, the Netherlands, 111 pp. Rijtema, P. E. 1958.Calculation methods of potential evapotranspiration. Rept. on the Conf. on Supplemental Irrigation, Comm. V1 Intl. Soc. Soil Sci. Copenhagen, June 30–July 4.

75. Robb, D. C. N. (1966). Consumptive use estimates from solar radiation and air temperature. Proc. Methods for Estimating Evapotranspiration, ASCE Irrig. Drain. Spec. Conf.; Las Vegas, NV, 169–191.

76. Shantz, H. L.,Piemeizel, L. N. (1927).The water requirement of plants at Akron, Colorado. J. Agr. Res. *34(12),* 1093–1190.

77. Shuttleworth, W. J.Towards one-step estimation of crop water requirements. Trans. of the ASABE, (2006). 925–935.

78. Shuttleworth, W. J.Wallace, J. S. (2009).Calculating the water requirements of irrigated crops in Australia using the Matt-Shuttleworth approach. Trans. of the ASBE, *56(6),* 1895–1906.

79. Smith, M., Allen, R.,Monteith, J. Perrier, A., Pereira, L.,Segeren, A. (1991). Report of the expert consultation on procedures for revision of FAO guidelines for prediction of crop water requirements.UN-FAO, Rome, Italy, 54 p.

80. Soil Sci. Society of America meeting, Aug. 20–24, (1962). Ithaca, NY.

81. Tanner, C. B. (1966). Comparison of energy balance and mass transport methods for measuring evaporation. 45–48, INJensen (ed). (1966). Proc. Evapotranspiration and its Role in Water resources Management. Am. Soc. Agr. Engr.; Chicago, Dec. 5–6, 68 p.

82. Teele, R. P. (1904). Irrigation and drainage investigations of the Office of the Experiment Stations, USDA.

83. Thoreson, Bryan, (2009). Personal information. David's Engineering.

84. Thornthwaite, C. W.(1948). An approach toward a rational classification of climate. Geographical Review, *38,* 55–94.

85. Turc, L.(1961). Evaluation des besoins en eau d=irrigation, evapotranspiration potentialle, formuleclimatiquesimplificeetmize a jour. (English: Estimation of irrigation water requirements, potential evapotranspiration: A simple climatic formula evolved up todate). Ann. Agron. *12,* 13–49.

86. USDA, (1970). Soil Conservation Service.Irrigation Water Requirements.Tech. Release No.21, (rev.), 92 pages.

87. Van Bavel, C. H. M. (1966a). Potential evaporation: The combination concept and its experimental verification. Water Resources Res.; *2(3),* 455–467.

88. Van Bavel, C. H. M. (1966b). Combination (Penman type) methods. *48,* INJensen (ed). (1966). Proc. Evapotranspiration and its Role in Water resources Management. Am. Soc. Agr. Engr.; Chicago, Dec. 5–6, 68 p.

89. Van Wijk, W. R.,de Vries,D. A. (1954). Evapotranspiration. Neth. J. Agr. Sci. *2,* 105–119.

90. Vitkevich, (1958). Determining evaporation from the soil surface. USDC PST Cat.310, (Trans. from Russian).

91. Walter, I. A., Allen, R. G., Elliott, R., Jensen, M. E.,Itenfisu, D.,Mecham, B., Howell, T. A., Snyder, R., Brown, P.,Eching, S.,Spofford, T.,Hattendorf, M., Cuenca, R. H., Wright, J. L.,Martin,D. (2000). ASCE's standardized reference evapotranspiration equation. Proc. FourthNat'l.Irrig. Symp.; ASAE, Phoenix, AZ.

92. Widstoe, J. A.(1909). Irrigation investigations: Factors influencing evaporation and transpiration. Utah Agr. Exp. Bull. *105,* 64 p.

93. Widstoe, J. A.(1912). The production of dry matter with different quantities of water. Utah Agr. Exp. Stn. Bull. *116,* 64 p.

94. Widstoe, J. A.,McLaughlin. (1912). The movement of water in irrigated soils. Utah Agr. Exp. Stn. Bull. 115.

95. Wright, J. L.(1982). New evapotranspiration crop coefficients. J. Irrig. Drain. Div. *108(IR1)*, 57–74.
96. McEwen, G. F. (1930).Results of evaporation studies conducted at the Scripps Institute of Oceanography and the California Institute of Technology.Bull. of the Scripps Inst. of Oceanography, Technical Series, *2(11)*, 401–415.
97. Richardson, B. (1931).Evaporation as a function of insolation. Trans. Am.Soc. Civil Engr. *99*, 996–1019.
98. Romanoff, V. V. (1956). Evaporation computations by simplified thermal balance method. Trudy CGI.

CHAPTER 2

EVAPOTRANSPIRATION WITH DISTANT WEATHER STATIONS

MOHAMMAD NABILEL-NESR and ABDURRAHMAN ALI ALAZBA

CONTENTS

Reprinted with permission from Mohammad N. El-Nesr and A. A. Alazba, "Evapotranspiration with Distant Weather Stations: Saudi Arabia," Chapter 14, pp. 337–348. In *Evapotranspiration: Principles and Applications for Water Management*, Megh R. Goyal and Eric W. Harmsen (Editors), (New Jersey, USA: Apple Academic Press Inc., 2014).

2.1   INTRODUCTION

Determining the evapotranspiration (ET) is the base of many disciplines including the irrigation system design, irrigation scheduling and hydrologic and drainage studies [12]. Perfect determination of ET is a big challenge for investigators especially in arid and hyper-arid regions. Actual crop ET is computed by multiplying reference ET by the crop factor. Reference ET is the summation of evaporation and transpiration produced by a reference crop in specific growth conditions (height, coverage and health). ET value depends on two main factors: the selected reference crop and the climatic data [4]. This chapter discusses the possibility of using the weather data recorded at these stations instead of reference agro-climatic data. Hence, the objective of this chapter is to estimate the ET values for the reference weather station (RWS) and the non-reference weather station (NRWS) at Riyadh – Saudi Arabia; and to compare these ET values with the reference evapotranspiration based on lysimeter observations for reference conditions.

The Kingdom of Saudi Arabia (KSA) is one of the most arid countries in the world and suffers persistent water shortage problems and more than 88% of water consumption in KSA is due to agricultural related activities [10]. Hence, several researches have been performed to assess the ET in KSA. Researchers have estimated the reference ET [2, 9]; and assessed the ET for specified crops [1, 7]; and have determined crop coefficients.

For open field agriculture, the reference ET has traditionally been predicted by using either grass (*ETo*) or alfalfa (*ETr*). Each of these two crops use some conditions to be considered as a reference crop [14]. The selection of either crop as reference crop has been studied by several investigators [6, 11, 13, 14]. It was recommended by the American Society of Civil Engineers Task Committee (ASCE-TC) to use a single equation for both reference crops, each with different constants [5]. They recommended standardizing the equation with two surfaces, the short crop (about 0.12 m height e.g., the clipped cool season grass) and the tall crop (about 0.50 m height e.g., the full cover alfalfa). The heights of crop, however, may vary according to the crop variety and location's geography. When using crop with different height, one should clearly mention the used height beside the ET data.

Climate data are acquired from Weather Stations (WS) whose location is an important consideration for the quality of the data. RWS should be located inside a cropped area (normally with grass) in order to ensure the same environmental conditions for station's gauges as that of the cultivated crops. On the other hand, stations located in these reference conditions usually record less temperature than Non-Reference (NR) weather stations [3]. This was attributed to the cooling effect of the crop. Allen [3] suggested an adjustment for the recorded temperature in NRWS so that the resultant temperatures could be used to give the reference ETo.

In many locations, RWS are not found especially for newly reclaimed desert areas. To perform preliminary studies for an area, one should use the data from nearest station. This situation is probably affected by the distance between the field and the weather station. There are 13 districts in KSA and some of these are larger in size than many countries. Arriyadh District's area, for instance, is 380,000 Km^2, which is 17% of the geographical area of KSA. The main weather stations in Arriyadh and other places in the kingdom are situated at airports.

2.2 MATHEMATICAL MODELS FOR ESTIMATING EVAPOTRANSPIRATION

The Eqs. (1)–(13) describe the selected theoretical models to estimate ET that were used in this chapter. For a complete list of all models, the reader should refer Chapter 2 in this book.

The original Penman-Monteith (PM) equation for determining the evapotranspiration [4] is given in Eq. (1). The Chapters 2 and 3 of the publication by Allen [4] describe in detail all the variables in Eq. (1). The main component of ET calculation is the air Temperature (T). Although it does not appear explicitly in the Eq. (1), yet it is included in most of the parameters: Δ, Rn, c_p, ρ_a, e_s, e_a, λ, γ, r_s, r_a, and G. Allen [3] concluded that correcting the temperature values of the nonreference weather stations [NRWS] to an adjusted value fixes the entire ET equation to give an acceptable value close to the reference weather station [RWS] value.

$$ET_O = \frac{0.408\Delta\left(R_n - G\right) + \gamma\left(\dfrac{900}{T+273}\right)u_2\left(e_s - e_a\right)}{\Delta + \gamma\left(1 + 0.34u_2\right)} \tag{1}$$

where: Δ is a slope of the vapor pressure curve ($kPa°C^{-1}$), R_n is net radiation ($MJm^{-2}day^{-1}$), G is soil heat flux density ($MJm^{-2}day^{-1}$), γ is psychrometric constant ($kPa°C^{-1}$), T is mean daily air temperature at 2 m height (°C), u_2 is wind speed at 2 m height ($m\ s^{-1}$), e_s is the saturated vapor pressure and e_a is the actual vapor pressure (kPa). The Eq. (1) applies specifically to a hypothetical reference crop with an assumed crop height of 0.12 m, a fixed surface resistance of 70 sec m^{-1} and an albedo of 0.23.

$$e_a = 0.005\left(RH_x e_o\left[T_n\right] + RH_n e_o\left[T_x\right]\right) \tag{2}$$

where: T_a = Minimum dry bulb air temperature (°C); T_x = Maximum dry bulb air temperature (°C); RH_x = Maximum relative humidity (%); and RH_a = Minimum relative humidity (%).

$$e_a = 0.005RH_a\left(e_a\left[T_n\right] + e_o\left[T_x\right]\right) \tag{3}$$

$$T_d = \frac{116.91 + 237.3\ln\left(e_a\right)}{16.78 - \ln\left(e_a\right)} \tag{4}$$

$$\Delta T = T_n - T_d \tag{5}$$

$$ET_{HG} = 0.0023 R_a \left(T_a + 17.8\right)\left(T_x - T_n\right)^{0.5} \tag{6}$$

$$ET_{PM} = c_1 + c_2 ET_{HG} \tag{7}$$

$$R_a = 37.6 d_r \left(\omega_s \sin \varphi \sin \delta + \sin \omega_s \cos \varphi \cos \delta\right) \tag{8}$$

In Eq. (8): d_r, δ, and ω_s are defined in Eqs. (9)–(11), and d_r is a relative distance from earth to sun, given by Eq. (9):

$$d_r = 1 + 0.033 \cos\left(0.0172 J\right) \tag{9}$$

Solar declination in radians is given by Eq. (10):

$$\delta = 0.409 \sin(0.0172 J - 1.39) \tag{10}$$

Sunset hour angle in radians is determined by Eq. (11), with φ = latitude in radians:

$$\omega_s = \arccos(-\tan \varphi \tan \delta) \tag{11}$$

Julian day (J) is calculated with Eq. (1), with M = Month of the year and D = Day of the month. J ranges from 1 to 366 (366 is for leap year).

$$J = \text{int}\left(\frac{275}{9} M + D - 30\right),$$

"if (M>2): For leap year, subtract 1; and for not leap year, subtract 2." (12)

Allen's method is summarized in Eqs. (1)–(5). The actual vapor pressure is calculated with Eqs. (2) and (3). The dew point temperature is determined with Eq. (4), in the absence of measured values. The temperature difference is determined with Eq. (5), where T_n is a minimum temperature. For arid and semiarid environments, if $\Delta T > 2$ then the maximum temperature (T_x) is adjusted to: $T_x{}^{(corr)} = (T_x - 0.5 (\Delta T - 2))$, where (corr) = Refer to the corrected value. Finally doing the same for T_n: if $\Delta T \leq 2$, then no correction is needed.

For sites with limited weather data, Allen [4] suggested using a modified version of the Hargreaves equation (HG) as an alternative method for determining ET. He also suggested calibrating the HG equation (Eq. (6)) using linear regression equation (Eq. (7) using the corresponding values from the PM Eq. (1). The units are: MJ-m^{-2} for extraterrestrial radiation in Eq. (8); MJ-m^{-1}-h^{-2} for ET_{HG} in Eq. (6); and mm-day^{-1}. The parameters in Eq. (6) were estimated with Eqs. (8)–(12).

2.3   WEATHER DATA SOURCES

Two types of data were used in this study: weather data to estimate the ET value; the field data to determine ET experimentally. For weather data, two weather stations (Campbell and Davis) at the educational farm of the King Saud University were selected as agro-climatic RWS [reference weather station]. Also two weather stations at Old Riyadh airport and King Khaled airport (Fig. 2.1) were used as domestic NRWS (nonreference weather stations). Reference field data was obtained from Al-Amoud et al. [1] based on five years project of ET evaluation through lysimeter studies in 9 zones throughout the country.

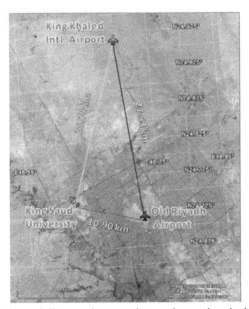

FIGURE 2.1   Location and distances between the weather stations in this research.

All of the weather data were recorded on daily bases while the field data was recorded on monthly bases. Hence, daily ET values for all the studied weather stations were calculated and later, the data was summarized as average monthly ET values. The recorded dataset varies from station to station. For airport's weather stations, complete records from 1985 to 2009 were obtained. For Campbell and Davis weather stations, the records were from 1993 to 2006. This research was limited to the least-size dataset, i.e., Campbell's dataset for an appropriate comparison. The databases of the studied weather stations were not so coincident. For all stations, the commonly available data parameters included: the dry bulb temperature (max., min. and avg.), relative humidity (max., min. and avg.), rainfall and wind speed (average). In addition to the common variables for airports stations, the wet bulb temperature (max., min. and avg.), the atmospheric pressure at sea level and at station level and the actual vapor pressure were also recorded. While for Campbell station, the actual vapor pressure and solar radiation are recorded. Finally, for Davis station, the only addition to common parameters was the solar radiation. This information is summarized in Table 1. Solar Radiation (Rs) and vapor pressure (ea) are essential parameters for computing ET. If not recorded at the weather station, these parameters were calculated using Eqs. (2)–(5) and Eqs. (8)–(12).

As mentioned above, the field data were obtained from Al-Amoud et al. [1]. The five years project used Alfalfa cultivated in weighing lysimeters located at Riyadh and at 8 more locations in the KSA. The daily and monthly values of irrigation, drainage, precipitation and water consumption were recorded for these stations. However, only the monthly results were available in the published research.

2.4   METHODS AND MATERIAL

For each of the four stations mentioned in Table 2.1, weather data were observed as daily records. Using the raw data, the researchers calculated ET_{PMg}, ET_{PMa} and ET_{HG}; where the suffixes PMg and PMa stand for Penman Monteith formula for 0.12 m grass reference crop and 0.25 m alfalfa reference crop, respectively.

The entire calculations were repeated after applying Allen [3] corrections to the temperature data but only for non agro-climatic stations. To simplify data representation and discussion, the symbol and numerical value were assigned to each data source as the shown in Table 2.2. Since the published data by Al-Amoud et al. [1] were on monthly bases and Allen [4] suggested calibrating Hargreaves formula using monthly data, subsequently, the research team converted the daily calculated data to the data on monthly bases.

2.5   RESULTS AND DISCUSSION

The ET data for the six datasets are presented graphically in Fig. 2.2. The charts are denoted by letters 'a' to 'f' for Campbell reference WS, Davis reference WS, old Riyadh airport corrected dataset, old Riyadh airport raw dataset, King Khalid airport corrected dataset and King Khalid airport raw dataset, respectively. Four ET values were plotted for each dataset: measured ET, {Px(L)}; grass based PM evapotranspiration {g}; alfalfa based PM evapotranspiration {a}; and Hargreaves method ET.

All of the calculated data groups were compared with the measured dataset and the correlation coefficient for each data group pair was calculated. The correlation coefficients are illustrated in Fig. 2.3. Then, the regression coefficients, mentioned in Eq. (7), were calculated between HG and PM formulas (*see* Table 2.3). Finally, the ET ratio between alfalfa and grass was calculated and compared to the value of 1.15 that has been reported by Doorenbos and Pruitt [8]. Table 2.4 shows the regression coefficients for the linear relationships between ET alfalfa and ET grass ET in this study.

TABLE 2.1   The Climatic Parameters That Were Recorded At the Weather Stations

Parameter	Weather station			
	Campbell	Davis	Riyadh old airport	King Khaled Int. airport
Dry bulb temp.	yes	yes	yes	yes
Wet bulb, relative	no	no	yes	yes
Wind humidity	yes	yes	yes	yes
Solar speed	yes	yes	yes	yes
Vapor radiation	yes	no	no	no
Atm. pressure	yes	yes	no	no
Commutative sea level	no	no	yes	yes
Commutative at station	no	no	yes	yes
Rainfall	yes	yes	yes	yes

TABLE 2.2   The Data Source For the Research and Identification

Variable	Location and name						
	Data is measured: Educational Farm, KSU			Data is calculated Airport			
	Project data	Campbell	Davis	Riyadh old airport		King Khaled Int. airport	
	0	1	2	3	4	5	6
Reference data	no	yes	yes	no	no	no	no
Corrected data	No need	No need	No need	yes	no	yes	no
Symbols	Px(L)	Cs (a, g, H)	Ds (a, g, H)	Oc (a, g, H)	On (a, g, H)	Kc (a, g, H)	Kn (a, g, H)
Longitude: 24N	44′12.24″	44′12.24″	44′12.24″	42′35.46″	42′35.46″	57′27.00″	57′
Latitude: 46E	37′ 14.90″	37′ 14.90″	37′ 14.90″	43′ 30.54″	43′ 30.54″	41′ 55.54″	27.00″

Symbols: A = Alfaalfa; C = Corrected, C = Campbell; D = Davis; g = grass; H = Hargreaves; K = King Khaled airport; L = Lysimeters; n = Normal; O = Old airport; P = Project data; s = Reference; × = Experimental.

TABLE 2.3 Linear Regression Analysis For ET Values Between Hargreaves [HS] and Penman-Monteith [PM] Based on Equations in This Chapter.

	Grass			Alfalfa	
	Equation	r^2		Equation	r^2
Cs	PMg = 1.280 × HG + 0.324	0.998	Cs	PMa = 1.608 × HG + 0.526	0.999
Ds	PMg = 1.563 × HG – 0.122	0.959	Ds	PMa = 1.993 × HG – 0.046	0.975
Kc	PMg = 1.189 × HG – 0.140	0.979	Kc	PMa = 1.481 × HG – 0.140	0.987
Kn	PMg = 1.296 × HG + 0.013	0.986	Kn	PMa = 1.619 × HG + 0.109	0.991
Oc	PMg = 1.214 × HG – 0.211	0.981	Oc	PMa = 1.517 × HG – 0.243	0.988
On	PMg = 1.298 × HG – 0.021	0.991	On	PMa = 1.624 × HG + 0.055	0.994

TABLE 2.4 The Regression Coefficients For the Linear Relationships **Between ET-Alfalfa,** ET-Grass and ET in this Study: Y = mX + C

	Cs	Ds	Kc	Kn	Oc	On
Slope, m	1.255	1.269	1.243	1.247	1.246	1.249
Intercept, C	0.125	0.158	0.053	0.107	0.040	0.092
r^2	0.999	0.999	0.999	0.999	0.999	0.999

Campbell and Davis weather stations are located at the educational farm of the King Saud University; the lysimeters' experiment of Al-Amoud et al. [1] was held at the same location. Figures 2.2a and 2.2b show the results of measured and calculated ET by the three mentioned methods. Both stations show underestimation of Hargreaves formula and overestimation of PM alfalfa, 'a,' calculations. It is strange that the grass ET, 'g,' almost coincides with the measured alfalfa data. This may be attributed to some lack of precision either in the field measurements devices or to some calibration errors of the weather stations. The data at old airport (Figs. 2.2c and 2.2d) behaves differently and the closer values to the measured alfalfa ET are the calculated alfalfa values, especially at months 1–3 and 9–12. The situation is different for King Khalid airport's station (Figs. 2.2e and 2.2f) as the raw data appear to give fuzzy trend dissimilar to the measured data, Fig. 1.2f. After applying the data correction, the shape of the curve improved dramatically, Fig. 2.2e. This is confirmed in Fig. 2.3, which shows the correlation coefficient (CC) between measured data versus each data group. Although all the values of CC are more than 0.9, which is a very good value, however, the 'Kn' dataset is the worst representation of actual state. On the other hand, it strangely appears that the corrected values of King Khalid's airport (K) are the top most accurate representatives of the measured data. This is probably due to the geographic condition of the 'K' airport, which is outside of the city and almost surrounded with desert lands, in addition to the long distance between the educational farm stations (EF) and the 'K' airport (about 25.7 km), as shown in Fig. 2.1. The old airport (O) is near (10.9 km), in fact almost in the middle of the city and surrounded by buildings, roads and some green areas. The correction of the 'O' data improves the 'g' and 'a' data groups, while it worsens the 'H' data group, as shown in Fig. 2.3.

From Fig. 2.3, it can be concluded that the PM calculations improve dramatically after applying the Allen [3] correcting algorithm to the data, while for HG formula, applying the corrections improves the accuracy for 'K' station but worsens it for 'O' station.

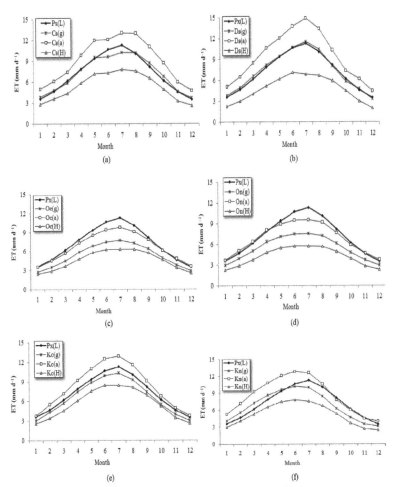

FIGURE 2.2 Monthly evapotranspiration of the studied region, showing three datasets in each chart compared to measured evapotranspiration, where, *a*: Alfalfa; *c*: Corrected; C: Campbell; D: Davis; g: Grass; H: Hargreaves; K: King Khalid airport; L: Lysimeters; n: Normal; O: Old airport; P: Project data; s: Reference; *x*: Experimental.

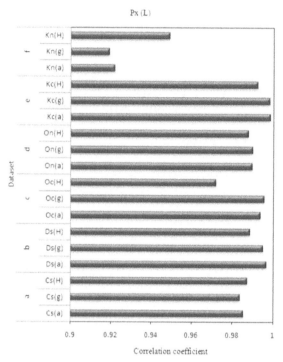

FIGURE 2.3 The correlation coefficients between the measured data and the calculated data for all data groups in the six datasets.

In general, Hargreaves equation gives very satisfactory results of ET for Riyadh city and the equation can be used trustfully especially in the absence of some climatic factors like wind speed and radiation. However, we applied the linear correction equations and found some excellent fitted equations, as listed in Table 2.2. All the equations are excellently fitted with minimum value of coefficient of determination (r^2) of 0.974. It can be approximated that $PM^g = (1.30?HG - 0.05)$ for grass reference crop, while $PM_a = (1.64?HG + 0.1)$ for alfalfa reference crop. For more accurate values each station should be calibrated, as shown in Table 1.3. The ratio between alfalfa ET and grass ET, (ETalfalfa/ETgrass, or ET_r/ET_o), is always taken as 1.15 for arid regions, as recommended by Doorenbos and Pruitt [8]. This value has been used by many researches [2] when studying the Saudi Arabia ET, We evaluated this value for each of the studied data groups and a linear relationship between ETalfalfa and ETgrass was obtained, as shown in Table 2.4. The slope of ET_r/ET_o is almost 1.25 for all stations and using less value may result in some bias in data.

SUMMARY

Due to the easiness of finding nonreference agro-climatic WS than the agro-climatic ones in the newly reclaimed areas, their weather data had to be corrected through simple procedure. Two reference and two nonreference WS were taken in Riyadh city and corrections were applied to nonreference WS only. We calculated the evapotranspiration using Penman Monteith and Hargreaves formulas. PM was calculated for two reference crops, i.e., alfalfa and grass. Calculated data were compared with measured data. Results show an admirable enhancement in data accuracy after applying the data correction to the nonreference stations. The simple ET formula of Hargreaves underestimates the actual ET. The situation changes after applying the simple linear fitting equation to the resulted values. The ratio between alfalfa and grass ET was found to be 1.25 for Riyadh area. It is concluded to use the temperature correction method when using nonreference stations. Hargreaves formula is recommended to be used after applying the suggested fit in this study, especially when the wind speed and radiation data are missing.

Reference agro-climatic weather stations (WS) are rarely found in newly reclaimed areas. The usage of weather data from nonreference WS may lead to inaccurate estimations of evapotranspiration (ET), especially if the nonreference stations are distant from the reclaimed location. Weather data from four WS located at Riyadh were used to calculate ET by using Penman Monteith (PM) and Hargreaves equations. PM equation was applied with both alfalfa and grass reference crops. Calculations were done with and without temperature correction for nonreference weather stations. All calculations were compared with measured lysimeter data and corrections in Hargreaves formula were suggested.

Results: (i) Weather data from nonreference WS can be used safely to calculate ET only when temperature corrections are applied. (ii) Hargreaves formula underestimates ET at all locations in the study area. By applying the simple linear correction to the data, highly acceptable results are obtained. (iii) The ET ratio between alfalfa and grass in Riyadh is 1.25. The study concludes that temperature correction for nonreference WS is essential to ensure acceptable ET calculations. Usage of Hargreaves formula is recommended with the corrections suggested in the study due to its simplicity.

KEYWORDS

- alfalfa
- arid Regions
- climatic data
- climatic stations integrity

- **data correction**
- **dew point temperature**
- **evapotranspiration**
- **extraterrestrial radiation**
- **grass**
- **Hargreaves method**
- **Lysimeter**
- **nonreference weather station**
- **Penman-Monteith method**
- **reference crop**
- **regression analysis**
- **vapor pressure**
- **weather station**
- **nonreference weather station**
- **reference weather station**

REFERENCES

1. Al-Amoud, A., Al-Takhis, A., Awad, F., Al AbdelKader A., Al-Mushelih, (2010). *A guide for Evaluating Crop Water Requirements in the Kingdom of Saudi Arabia.* King Abdulaziz City for Science and Technology, Abdulaziz – KSA, 104.
2. Al-Ghobari, H. M. (2000). Estimation of reference evapotranspiration for southern region of Saudi Arabia. Irrig. Sci. *19,* 81–86.
3. Allen, R. G. (1996). Assessing integrity of weather data for reference evapotranspiration estimation. J., Irrig. Drainage Eng. *122,* 97–106.
4. Allen, R. G. (1998). *Crop Evapotranspiration: Guidelines for Computing Crop Water Requirements.* First edition by Food and Agriculture Organization of the United Nations, Rome. 300.
5. Allen, R. G. (2005). *The ASCE Standardized Reference Evapotranspiration Equation.* ASCE Publications, Reston, VA. 216.
6. Allen, R. G., Walter, I. A., Elliott, R., Mecham, B., Jensen, M. E. (2000). Issues, requirements and challenges in selecting and specifying a standardized ET equation. CMIS.
7. Al-Omran, A. M., Mohammed, F. S., Al-Ghobari, H. M., Alazba, A. A., (2004). Determination of evapotranspiration of tomato and squash using lysimeters in central Saudi Arabia. Intl. Agric. Engr. J. *13,* 27–36.
8. Doorenbos, J., Pruitt, W., (1977). *Guidelines for Predicting Crop Water Requirements.* Food and Agriculture Organization of the United Nations, Rome. 144.
9. El-Nesr, M., Alazba A., Abu-Zreig, M., (2010). Analysis of evapotranspiration variability and trends in the Arabian Peninsula. Am. J. Environ. Sci. *6,* 535–547.

10. Faruqui, N. I., Biswas, A. K., Bino, M. J., (2001). Water Management in Islam. United Nations University Press, Tokyo. 149.

11. Howell, T. A., Evett, S. R., Schneider, A. D., Dusek, D. A., Copeland, K. S., (2000). Irrigated fescue grass ET compared with calculated reference grass ET. Proceedings of the Decennial Symposium, Nov. 14–16, American Society of Agricultural Engineers, Phoenix, AZ, 228–242.

12. Irmak, S., Haman, D. Z. (2003). Evapotranspiration: Potential or Reference. University of Florida.

13. Wright, J. L. (1996). Derivation of Alfalfa and Grass Reference Evapotranspiration. In: *Evapotranspiration and Irrigation Scheduling*, Camp, C. R., Sadler, E. J. (Eds). American Society of Agricultural Engineers, Michigan. 133–140.

14. Wright, J. L., Allen, R. G., Howell, T. A. (2000). Conversion between evapotranspiration REFERENCES and methods. Proceedings of the 4th Decennial Symposium, NWISRL Publisher, USA, AZ-Phoenix. 251–259.

CHAPTER 3

PRINCIPLES OF DRIP/MICRO OR TRICKLE IRRIGATION

R. K. SIVANAPPAN and O. PADMAKUMARI

CONTENTS

This chapter is modified from R. K. Sivanappan and O. Padmakumari, 1980. Drip Irrigation. Bulletin by TNAU, Coimbatore, India.

In this chapter, conversion rate for an Indian currency is Rs. 64.00 = US$ 1.00 on June 30, 2014. The prices of drip irrigation system/crops are based on 1979–1980 prices. Information in this chapter is based on research studies by the authors and interviews with farmers/scientists in India. Authors acknowledge the support by College of Agricultural Engineering and Water Technology Center of TNAU, Coimbatore; and by Indian Council of Agricultural Research (ICAR), New Delhi, India.

3.1   INTRODUCTION

From the time immemorial, gravity irrigation by way of flooding in furrows and various forms of basins has been practiced all over the world. In surface irrigation, water infiltrates into the soil while traversing and also while standing in the furrows, borders or basins. The depth of infiltration depends on the quantity, duration and rate of stream flow, the gradient, soil texture and structure. The overall irrigation efficiency in surface irrigation varies from 25 to 60%, and is considered low because of water loss due to seepage, evaporation and deep percolation. Additionally, it may cause erosion, salination and water logging. Because of low irrigation efficiency, larger quantities of water are required to carry it from the source to field, calling for greater storage facilities and channel capacities, larger structures and extensive drainage systems.

In gravity irrigation, about 5 to 10% of the cultivable area is taken up by irrigation and drainage channels. Moreover, the water wasted by the use of inefficient methods restricts the area that can be cultivated with a given amount of water, thus inflating project costs per unit area, reducing overall project returns, and impairing project feasibility. The preparation of land for gravity irrigation is costly and time consuming, in addition to the skills and experiences of local conditions that are required. Losses in productivity through the removal of the upper fertile layers are considerable, even though only temporary in character. The farmers are unable to afford investment in other supporting agricultural inputs, necessary for full utility of the project. Compared to modern pressure irrigation methods (Fig. 3.1), gravity irrigation has following disadvantages:

- more water is needed per unit area;
- danger of accumulation of water in the subsoil causing water logging, salinity or alkalinity;
- costly and time-consuming, preparation of land including careful surveying, scraping and leveling;
- need for care in the application of irrigation water.

Another irrigation method is sprinkler irrigation, which is almost like rain that can be controlled both in time and intensity. In this method, water is delivered under pressure into a system of portable, lightweight pipe lines with sprinklers mounted at regular intervals. Complete sprinkler irrigation systems are now available to suit almost all crops and a wide range of conditions. When properly designed and all its components selected in accordance with wind conditions, cropping patterns and soil water properties, the sprinkler system can be operated at high irrigation efficiencies of 65–80% compared with gravity irrigation. The water logging, salination and deep percolation can be eliminated, if the system is properly designed and installed.

During the last 70 years, sprinkler irrigation has undergone great improvements. Sprinklers are being constantly modified; their construction and hydraulic performance have been improved; and quick-coupling light pipe is now stronger and less corrosive. The couplers are made to operate more reliably under a wide range of pressures, and pumps are engineered to suit the hydraulic characteristics of the system. It is possible today to order complete sprinkler irrigation systems to suit almost all crops and a wide range of conditions. However, one should not oversimplify the acquisition of the sprinkler system, though much can be done to standardize components and prepare improved design criteria [7].

During 1960s, agricultural development in Israel was particularly relevant in this context, since almost 90% the cultivated lands were sprinkler irrigated. Due to high irrigation efficiency, these systems were adopted due to water scarcity and unfavorable distribution of the water resources. Farm water utilization has been optimized due to concurrently applied advanced methods of cultivation, intensive research in plant water use, and a concentrated extension programs. Therefore, crop yield has increased with reduction in water application. Proper and controlled use of water has minimized the need for drainage and reduced salinity hazards and leaching requirements, even though the farmland in India is mainly arid to semiarid and the water quality is often affected by excessive salts [8, 9].

The sprinkler irrigation can be operated at an irrigation efficiency exceeding 80%, when the system is properly designed and all its components are selected in accordance with wind conditions, cropping patterns and soil-water properties. Consequently, water use per unit land is lower in sprinkler irrigation than in the gravity irrigation systems [16, 19]. Excess irrigation leading to waterlogging and salination may be almost entirely avoided with sprinkler irrigation. Deep percolation can be eliminated, unless necessary for leaching of salts. Irrigation can be easily timed according to plant requirements, soil texture and root depth of crop. Thus, shallow rooted crops can be given light and frequent applications. With high uniformity of distribution, advantageous environmental crop conditions are favorable.

As time goes on, more and more land will have to be cultivated and irrigated to meet the needs of growing populations, as well as to replace lands which are being abandoned owing to salination and loss of fertility. New lands, often of shallow depth and poor topography, will have to be brought into production. These lands can only be irrigated efficiently under sprinkler and micro irrigation. Furthermore, such virgin or desert lands will invariably be situated far from water resources, requiring long conveyance lines and possibly high lifts. Under such conditions, economic considerations require maximum efficiency in water conveyance and application.

The main disadvantage of sprinkler irrigation is high initial investment. It requires a booster pump for the sprinklers to work. Another drawback is that saline

Young engineers restoring and developing the old Mughal irrigation system during the reign of the Mughal Emperor Bahadur Shah II.

Furrow Irrigation of sugar cane.

Irrigation of land in India.

Level basin flood irrigation on wheat.

FIGURE 3.1   Conventional irrigation systems.

FIGURE 3.2   Typical sprinkler irrigation

FIGURE 3.3   Typical micro irrigation system for vegetables (left) and vines (right).

water when applied through sprinklers, come in contact with leaf foliage and cause disease in some crops (Fig. 3.2).

During the last 40 years, advanced method of water application, called drip/trickle/daily/or micro irrigation, has spread throughout the world. It has been noticed that drip irrigation plays a considerable role in the national agricultural economy and at the same time help us to use limited water resources more profitably. In this system, water is delivered to each plant near the root zone, through a network of tubing. It can operate under low or medium pressure and only the required quantity of water is delivered daily in order to avoid pant water stress. The system design is based on the criteria that the minimum moisture stress which will be maintained on a substantial part of the root zone and this will be achieved at minimum capital and labor cost (Fig. 3.3). The main advantages of drip irrigation system are: Reduced labor cost, water saving, leveling of the land can be minimized, reduction of weed growth, increase in crop yield and better quality, faster maturity of young plants due to absence of moisture stress, reduced fertilizer requirement, reduced incidence diseases and pests, decreased tillage operations, feasibility of usage of poor quality water. The main disadvantages of drip method of irrigation are sensitivity of clogging of emitters, problem of uniform moisture distribution problem, and high initial cost of irrigation system [7].

The drip irrigation method has been successfully used for cultivation of strawberries, grape fruit, lemon, lime, orange, walnut, apple, tomato, cucumber, celery, potato, pepper, melon, corn, eggplant, pea, lettuce, ornamental trees and shrubs, bedding plants, bulbs, avocado nurseries, radish, apricot, plums, cherries, almond, sugarcane, banana, coffee, wheat, etc.

Industrial production of drip irrigation systems started as soon as serviceable emitters and suitable other components were developed. Initial field trials resulted in very high yields for various crops, which resulted in rapid acceptance by passing the normal process of research and development in countries like Israel, Australia, South Africa, and USA. Now, drip irrigation is an accepted technology among marginal and progressive farmers.

There has been tremendous increase in drip-irrigated area in the world. For the recent data on micro irrigation, the reader can refer to 'http://www.icid.org.' Since 1975, the area under drip method has been increasing every year and more crops are being drip irrigated in Israel, USA, Australia, South Africa, India, Middle East and other parts of the world.

3.2   BASICS OF DRIP IRRIGATION

Drip irrigation is an advanced method of applying water and fertilizer near the root zone of the crop with the help of low cost plastic/PE pipes and drippers. The dripper achieves a three dimensional differential spread of water maintaining low levels of soil water tension. Under highly controlled conditions of soil moisture, fertilization, salinity and pest control, the drip irrigation method has significant effect on crop response, timing of harvest and fruit quality.

In this system, water to the plant is supplied from a point source. The water supply is along the lateral pipe at uniform intervals according to the row spacing and crop type. The soil is saturated close to the point source with a gradual decrease in moisture content in all directions away from the source.

3.2.1 EFFECTS OF DRIP IRRIGATION ON CROP WATER USE

Water plays a vital role in all stages of plant growth. Nature has given each stomata of leaf two guard cells, which are capable of closing the pore to prevent a level of water loss that could cause permanent damage to the leaf. The time, at which the stomata will close on any particular day, will depend on the evaporation demand for that day and on the ability of the tree to extract water from the soil at the required rate. Between two irrigations, the plant is not able to extract water at a rate that will meet its maximum needs. Hence, it is better if the leaves produce sugar for a restricted number of hours each day rather than to provide a luxury supply for part of the time and drought at other times. The daily flow drip irrigation is based on this concept [3]. Therefore, with drip method, it is possible to provide daily maintenance of an adequate section of the root zone at field capacity during the growing and productive cycle. A rather high moisture regime prevails within the quite sharply defined boundaries of the wetted bulb, which enables the development of live roots. The development of root in the drip system is similar to other surface methods. The effectiveness of the system will depend to some extent on the area around each outlet, which can be maintained in a moist state.

Research results have revealed that in a permeable shallow clay loam over fine sandy clay subsoil, a single emitter with a flow rate of 2 to 10 lph has resulted in an irregular circular shaped wetted area of 3 to 4 meters diameter. Studies with 1.0 lph emitter with a dripper spacing of 60 cm have resulted in the wetted area around each emitter joining to form a continuous wetted band along the lateral line. In the shallow permeable soils over relatively impermeable subsoil, above emitter flow rates resulted in 50% to 70% of the root zone area that was wetted around each outlet. To maintain this area at a minimum soil moisture stress, a daily water application is made depending on the conditions of the previous day. Water depth is estimated as a proportion of the Class A pan evaporation and taking into account the estimated area of soil from which the plant draws its needs.

3.2.2 POTENTIAL TRANSPIRATION AND CONSUMPTIVE USE

In drip irrigation, it is possible to apply small quantities of water at a desired interval. Therefore, it is possible to maintain a low moisture tension within the root zone. The drip irrigation thus affords growing conditions with a unique moisture regime called potential transpiration. In order to understand and properly define potential transpiration, one must consider potential evapotranspiration (PET), which is defined as the consumptive use (CU) from a sodded field having unlimited water availability. There are many climatic models available to calculate PET and CU.

Enough research has been conducted to estimate PET [7, 9] by Penman, Pruitt, Christiansen, Blaney – Criddle, etc. Among these methods, the class A pan has been found to be simpler to determine the CU of a crop. A very high correlation has been found between ET and USDA Class A pan (Epan). The correlation of ET to Epan over weekly to monthly periods is relatively high [7, 9]. Penman [9] developed a seasonal relation to evapotranspiration for grass to pan evaporation. He established that PET/Epan ratio was 0.6 for November to February, 0.7 for March, April, September and October and 0.8 for May to August. Similar observations made by various scientists in different locations have shown that PET/Epan ratio varied from 0.6 to 0.8 under almost all conditions [9]. In drip irrigation, a very small area of the soil surface is wetted and is almost totally shaded by plant foliage. This reduces evaporation considerably. Moisture is thus being lost to the atmosphere mostly through transpiration process. Broadly it can be stated that: (i) Potential transpiration should in all cases be smaller than PET; (ii) Depending on the crop type, the quantity of water transpired will vary and this must be taken into account; and (iii) As the soil is at field capacity, the roots are constantly absorbing water without being subjected to any water stress.

3.2.3 DESCRIPTION OF DRIP IRRIGATION SYSTEM (FIG. 3.4)

The drip irrigation system consists of the following components:
- a head unit consisting of a riser valve, pressure gage.
- plastic main supply lines of required lengths and diameter of 3.75 to 7.00 cm to carry the desired discharge.
- lateral lines, of diameter 10 to 15 mm made of alkathene or polyethylene, are connected to the main line.
- emitters or drippers inserted on the lateral at desired spacing.
- a fertilizer tank filled with concentrated nutrient solution and connected to the head unit.

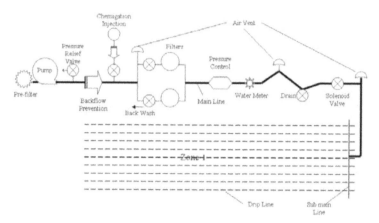

FIGURE 3.4 Typical layout of a drip irrigation system, showing drip line (laterals), submain, main, pump, valves and accessories.

Dripper spacing and size is selected according to row-to-row spacing, plant to plant spacing, crop water needs, and soil characteristics. The laterals are connected to the main line. The length and diameter of the laterals depend on the desired flow rate and economic considerations. Application rates can be adjusted by using suitable emitters with different flow rates.

During the plant growth, there are a number of irrigation cycles depending on the interval between water applications (Figs. 3.5 and 3.6). Each cycle consists of the following stages: At the time of irrigation, the soil water matric potential in the main root zone is zero. It gradually reaches the field capacity (33 cbars). Before the next irrigation, it can increase upto 100 to 200 cbars. At the same time, the osmotic potential increases depending upon the salt concentration of the irrigation water.

FIGURE 3.5   Surface drip irrigation.

FIGURE 3.6   Soil wetting pattern under inline emitters.

Increasing the discharge rate will result in enlarging the diameter of the wetted area and increasing the water content of the upper soil layer near the drippers. The low metric tension with drip irrigation allows the use of higher salinity levels or osmotic tensions without affecting the yield.

3.2.4   IRRIGATION INTERVALS

To evaluate the effects of irrigation frequency on tomato yield, research studies were conducted at the research farm of Tamil Nadu Agricultural University, Coimbatore a water scarcity district of Tamil Nadu State in India. Comparing the conventional irrigation system (control) with the drip irrigation method, it was noticed that it was practically impossible to irrigate at such short intervals in conventional irrigation. For tomato and okra crops, there was no significant reduction of yield when irrigation was given on alternate days. In the control method, the crop yield was decreased when water was applied at 75% available soil moisture depletion (ASMD). Table 3.1 shows effects of irrigation treatments on tomato yield.

A comparison was also made in drip irrigated bhindi (Okra: *Abelmoschusescul-entus Moench*) with water of two qualities. The electrical conductivity (EC) was 1.0 for the wetland water and 3.8 for the well water. The differences in bhindi (Okra) yield between the two water qualities were not significant (Table 3.1).

TABLE 3.1   Tomato and Okra Yield Under Drip Irrigation at TNAU, Coimbatore, India

Irrigation treatment	Yield	Water applied	Rain	EC
	Kg/ha	cm	cm	–
Tomato				
Daily: drip irrigation	14,000	13.5	40	–
Alternate, drip irrigation	14,100	13.5	40	–
Control, 50% ASMD depletion	13,500	60.00	40	–
Control, 75% ASMD	11,500	45.00	40	–
Drip irrigated Bhindi (Okra): with saline water				
Demonstration plot	10,450	10.5	5	1.1
Field near soil and water conservation block	10,500	12.0	5	3.8

Control: Conventional irrigation system; EC = Electrical conductivity, mmhos/cm

3.2.5   DRIP IRRIGATION VERSUS SOIL SALINITY

Salinity is an agricultural hazard. Under drip irrigation salts accumulate both at the periphery of the wetted soil zone and on the soil surface. The concentration of the soil solution influences greatly the physiological process and growth of a plant. Considerable research on soil salinity under desert conditions In Israel has been done to ascertain the salinity levels and to evolve proper strip cultivation practices. The quality of water is the prime cause of salinity. It depends on three main factors: (i) the sodium concentration and the ratio of Na ions to the concentration of calcium and Magnesium ions; (ii) the concentration of boron and other micro nutrients; and (iii) the total salt concentration measured by electrical conductivity (EC).

It has been observed that plant growth decreases as the osmatic tension of the soil increases. The use of saline water in drip irrigation resulted in better plant response and yield than with any other irrigation method. An understanding of salt distribution in the root zone and soil profile is essential. The initial salt distribution is consequent to the water flow and generally described as:

• A shallow pocket of accumulated salts, surrounding the "transmission zone," varies from 30 to 80 cm in radius, depending on the flow rate and irrigation depth.
• A relatively backed soil "cone" with a low salt concentration.
• A higher salt accumulation at deeper level located radially about the "wetting front." Overlapping webbed shapes might develop higher salt concentrations in the adjacent zone, but it shows no negative effect as long as higher moisture level is maintained.
• The dry space between lateral lines does not undergo any moisture depletion or salinity process.

By frequent irrigation, a state of balance is reached in drip system when salts can be kept in the periphery beyond the actual root zone. The salinity level in the root zone throughout the growth period affects crop response and influences the yield. Some soils may develop high salt concentration after one or two crop seasons. It requires periodical leaching. If adequate rain is not available, leaching may be done through drip system itself to provide surplus water. During leaching, it must be seen that back flushing or salt does not occur.

3.2.6   EFFECTS OF IRRIGATION METHODS

Drip irrigation method results in less chloride content in plant leaves compared with other irrigation methods. It has also been found that plants, which were irrigated by conventional sprinkler or furrow method, does not suffer any shock or other marked damage when they are brought under drip system. Drip irrigation experiment indicated a definite tendency for the roots to concentrate in the upper soil layer. The 90% of the total roots are within the first 30 to 40 cm depth for annual crops and 80 to 100 cm for perennial crops. Blank [29] compared the effects of irrigating only part of the root system in apple tree on yield production with conventional irrigation practices. When 25% of the roots were irrigated, 74% of the control yield was obtained. Watering 50% of the roots yielded 88% of the control and watering 75% of the roots gave 94% of the yield. In other words, irrigation only 50% of the root zone under conditions of low moisture tension allowed the plant to produce almost 100% of the yield obtained by normal practices. Goldberg found similar results in Arava desert and the coastal plains of Israel [4–6].

3.3   TYPES AND METHODS OF DRIP IRRIGATION

The Drip system consists of a main line, sub mains, laterals, a riser valve, vacuum breakers, pressure gages, water meters, filters, a fertilizer injector, flush valve and pressure regulator. From the main, feeder lines are run across the field and laterals are connected on the same. Low-density polyethylene pipes are laid along the plant or tree rows with drippers inserted at appropriate intervals. The emitters are designed to supply water at low rates (1 to 10 L/hour) directly to the soil, near the plant. Low pressure ranging from 1/3 to 2 Kg/cm^2 is sufficient to operate. Pressure control valves are also fitted. Usually 3.75 to 7.00 cm diameter tubes are used for main and 2.5 to 3.75 cm tubes are used commonly for submains in the drip system. For laterals 6, 9, 15, and 18 mm diameter pipes are used to suit various situations. In hilly areas where pressure variations become excessive, sub mains with control valves are used to feed laterals. A vacuum release valve must be installed at the highest location. Centrifugal pump can supply water at constant pressure and adjustable discharge. Once an even flow from each outlet is achieved, the system should not be altered and water control should be made based on the running time (irrigation duration). On heavy soils, one outlet per tree may be enough and for lighter soils two outlets per tree should be enough. Achieving regular flows over a large area is difficult, as changes made along the line affect the discharge from the outlets.

Twin-wall (biwall) chamber tubings are sometimes used for drip laterals. It consists of a main chamber and a secondary chamber. Water is discharged through orifices in the wall of main chamber into the secondary chamber, which then emits out water for irrigation. The ratio of number of orifices in the main to the secondary chamber is relatively small, because water is released under lower pressure than the pressure in the main chamber. The main advantages of the biwall drip tubing are that larger orifices can be used without increasing the discharge and longer lengths of tubing can be used, with a more uniform orifice discharge than single chamber tubing (Fig. 3.7).

The efficiency of any irrigation system can be broadly expressed in two ways: (i) application efficiency; and (ii) distribution efficiency. In the drip irrigation, the water is applied to the root zone and the surrounding areas are dry. Therefore, the application efficiency is high. When the emitter discharge variation is not greater than 10%, the distribution efficiency is also high [7]. This can be obtained by adjusting the inlet pressure and lateral length.

3.3.1 DRIP IRRIGATION METHODS OUTSIDE INDIA

In Western Australia, the types of drip irrigation systems are: Readymade out-fits; adjustable outlet system; Micro-tube system; and microsprinklers. The readymade system relieves the farmer of design and adjustment work. Adjustable system allows greater latitude in main and lateral design, but the effort involved in trial and error adjustment may outweigh this. The task is reduced with microtube system when changes in length are used to compensate for pressure variation and friction losses. However, microtube system must be designed for a particular location and crop and cannot be used in any other situations.

In Israel, most of the trickle irrigation systems are of permanent type and are not designed for moving from one place to other. However, field crop requires a portable system. In portable drip irrigation method, the operation is accomplished by attending to several rows with only one trickle lateral. After the irrigation of a given row has been completed, the drip lateral is moved manually to the next parallel row. There are various types of emitters (Fig. 3.7) that are used in Australia:

1. **Construction orifice emitters**: By constricting the water flow at very short and narrow outlets, considerable head losses are sustained and the rate of discharge is reduced accordingly.
2. **Vortex emitters:** Water is forced tangentially into a circular chamber and water path is formed into a whirl. The centrifugal force develops sufficient head losses.

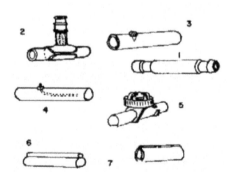

1. In-line emitters.	5. Compensating pressure emitter.
2. On-line emitters.	6. Low pressure emitter lines.
3. Micro sprinklers.	7. Biwall drip lines.
4. Micro-tube (spaghetti).	

In France, there are two types of drip irrigation systems:

Inline button drippers.

Inline vortex emitters with orifices.

Biwall PE tubing

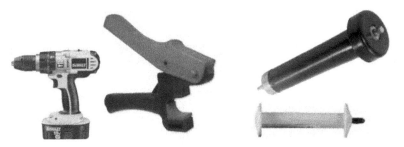

FIGURE 3.7   Types of emitters.

FIGURE 3.8   Manual hole puncher for drip tubing.

1. **Conventional trickle irrigation:** Water is distributed to outlets without atomization and without wetting the land. The density of watering point enables the subsoil to be moist while greater part of surface soil is dry. Water is delivered at low pressure of 1 to 2 atmospheres (Fig. 3.7).

2. **C.N.A.B.R.L. localized drip irrigation system**: In this system, manual hole puncher (Fig. 3.8) is used to make holes at desired locations on the PE pipes and a power drill is used to make holes on the on PVC pipes. These openings have a diameter from 1.6 to 2.1 mm, to function at a pressure between 0.4 and 1.5 atmospheres. The jet, which escapes through the holes, is deflected by a nozzle. The pipe with aperture is placed at the bottom of a trench and small heap of soil is made over the pipe along the length so that each hole is in a trough. The trench is not cultivated, but is grassed over, which does not affect orchard crops.

1. **Long path inline emitters**: Head losses are caused by longitudinal friction of the water flow along the walls of a microtube of relatively long path.

In England, drip irrigation system is mainly used for crops, which are cultivated in glass houses and film plastic structures. Trickle irrigation systems are used to provide water and nutrients for container grown plants standing on capillary sand beds. There are three following methods:

a. **Screw thread method**: In this method water flows through restricted passage of water around a helical groove as in a screw thread. Plastic nozzles are inserted into flexible PVC tubes. The nozzle should not be more than 24 m distance, from the inlet. A hole is drilled at a spacing of 30 cm and a nylon insert internally threaded is pushed in. A 6 or 8 way distributor head is then screwed-in into this at the same time making the joint water tight. Small bare PVC tubes are then run from the distribution head to a drip nozzle in each pot. For plants in beds, PVC tube is laid down on one side of plant and the nozzles are fitted at each location of the plant.

b. **Capillary tube method**: By the restricted passage of water through a small diameter capillary tube, water is let out. Tubes are 0.8 mm diameter and 85 cm long, and are connected to a 12.5 mm (0.5") PVC lateral running down the crop row. Drip lines may have tubes spaced greater than 75 mm (3"). The drip line is supplied by a resistance tube of diameter suitable to drip, with a pressure from 0.2 to 1.2 atmospheres in the drip line.

c. **Stitched polyethylene tube method**: The idea of using thin polyethylene (PE) tubing stitched with a synthetic fiber thread down one side to distribute water was introduced in England from USA in 1970s. The system is used for giving even watering to flower crops grown in beds. The "Ooze tubes" PE tubes are 16 mm (5/8") in diameter and one tube is provided for each row of plants. With 2.5 cm diameter PVC ring main around the bed and ooze tubes fed alternatively from each side by a 3 mm PVC tube, an even spread of water can be obtained. One tube of 1.3 m (4 ft.) long will give a flow rate of 200 mL/min.

d.   **Larger bore stitched tube** of 18 mm diameter (when inflated) is available in the United Kingdom, which is of Danish origin. This is suitable for crops grown in rows or beds. It can deliver an output of 250 mL/min. At present, it is not commonly used. This particular product called "**Seep Hose**" is manufactured by machine from PE sheet. The PE tube used for British ooze tube is not manufactured to narrow diameters and is unsuited to stitching on an automatic machine. Seep Hose is stitched on an automatic machine. They are not easy to correct for gradients.

Design of a suitable drip irrigation system under Indian conditions was evaluated using different types of tubes, at Tamil Nadu Agricultural University, Coimbatore. The success of any irrigation system depends greatly on its design, though it is less critical for gravity systems where reasonable performance can be achieved provided that the design is based on generally acceptable principles. Since drip irrigation system can result in crops of higher yield and better quality, design based on proper assumptions and reliable data is essential as well as proper materials for a sound irrigation network. During 1970s and 1980s, the following tubes were tried for use as drip lines suitable for Indian conditions:

- Alkathene pipes;
- White rigid PVC pipes;
- Krishi (Farmer) hose;
- Flexible plastic hose;
- Thin tube with polythene films.

From the research studies at Tamil Nadu Agriculture University – Coimbatore, it was observed that 12.5 mm black alkathene tubes are more suited and efficient because of low cost compared to other pipes. The 2.50 cm diameter PE hose, stitched with an ordinary sewing machine, seeps along its length with increase in pressure. For a pressure below 0.5 m, it is found to be not leaky but begins seeping through the stitch point at or above 0.5 m pressure head in the lateral pipe. As water is coming along, it can be used for closely spaced crops such as radish, sugarcane, Grain sorghum and other millets. In 1980, the cost of the stitched pipe was nearly Rs. 0.20 per meter. The researchers noticed that this PE tube can easily move from one place to other due to wind, thus a main disadvantage of PE tubes.

The white rigid PVC pipes are more durable than the alkathene pipes. However, the initial cost is higher than that of alkathene pipes and there is algae formation in the inner sides of lateral tubes. Therefore, it should not be used and laid on the soil surface in open fields for crops of long duration.

The Fig. 3.7 presents types of emitters that are currently used in drip irrigation. These days, self – flushing emitters are also available.

3.3.2   DRIPPER WITH A HOLE OF 1 MM DIAMETER WITH SOCKET

The water flow through the hole is similar to as through an orifice. A hole of 1 mm diameter can be punched on the lateral using a punching machine (Fig. 3.8). According to the pressure at the inlet, the water is discharged through the orifice, as a socket to dissipate the pressure of out-coming water. And the flow through the orifice or nozzle is always turbulent. The emission uniformity for this type of emitter is about 90% for a laminar flow at a constant pressure. The uniformity coefficient is calculated as follows:

$$EU = (Qn/Qa) \times 100 \qquad (1)$$

where: EU = Uniformity coefficient; Qn = Minimum emitter discharge; and Qa = Average discharge. The field evaluation of EU is made by taking water samples for a known time at three to five locations along four different lateral lines equally spaced throughout a representative area. The selected locations must also include the highest and lowest locations in the field. For further details, reader is referred to chapter 14 by Goyal [7]. Nomographs are available to check if EU is acceptable [7].

To evaluate the EU on a level field, 1.25 cm diameter alkathene pipes with 1 mm diameter holes and sockets were placed and fitted to 5 cm submain. For different pressures, the discharge through the orifice was recorded. For different pressures, the orifice discharge was found to be uniform for the lengths particular flow rate (Table 3.2). It is concluded that by using the hole and socket method for drip emitters, it is cheaper and it can be used satisfactorily for normal fields of flat topography.

3.3.3   MICRO-TUBE EMITTERS

These are in general most effective under Indian conditions, because of low cost and easy adjustment of flow. In these emitters, the flow can be adjusted even for very low flow rate due to minor variations for wide range of pressures. Uniform size and cross section of emitters is also available. Using microtube emitters in five locations in the field equally spaced, the discharges were measured for different pressures. The readings indicate that uniformity coefficient of 90–95% was possible with these outlets on a level ground.

TABLE 3.2   Distribution Efficiency of 12.5 mm Lateral Line with: An Average Flow of 0.5 L per Emitter, Lateral Length of 30 m, P = psi, and Hole Spacing of 60 cm

Maximum discharge, qmax	Minimum Discharge, qmin	Average discharge	Flow varia- tion, = qmax – qmin	Discharge variation	Distribution Efficiency, Ed
mL per emitter		mL	mL	%	%
500	460	480	40	8.0	91.97
530	490	510	40	7.5	92.16
500	450	476	50	10.0	89.5
520	480	500	40	7.1	92.0
530	500	571	30	5.8	94.73
520	470	495	50	9.6	89.9
535	497	516	38	7.1	92.7
528	510	519	18	3.4	96.6
521	501	511	20	3.8	96.1
495	480	487	15	3.0	96.6

Total flow (discharge) for all observations = Qtotal = 30,000 mL = 300 L.

Time of flow = 10 min for all observations.

The qmax and qmin are the values taken in the same lateral at different emitter locations; and the lines are selected at random for different locations on level ground.

Distribution efficiency, % = Ed = {[1 – q] ÷ [qavg]}?100.

where: q = variation in discharge, qavg = average discharge.

3.3.4   NOZZLES

In India two types of nozzles are made locally, which are suited as emitters. One nozzle is provided with threads and a hole is drilled to the other end. The nozzle is inserted in the lateral line by punching a hole. The disadvantage of this nozzle is that at high pressure, water comes out in the form of a sprinkler and does not wet the root zone around the plant. Second type of nozzle is more or less like the joint where the bend is provided for dissipating the energy. On the top is provided a cap which facilities cleaning. The emitter is suitable where water is also applied for leaching requirements.

3.4   WATER REQUIREMENTS FOR DRIP IRRIGATION SYSTEMS

Efficient use of drip irrigation system requires knowledge of soil moisture content. Gravimetric method is a common method to find soil moisture content. Crop water requirement depends on factors such as: temperature, climate, relative humidity, and plant/soil characteristics. The irrigation planner must have a general idea of the crop water requirement. The soil moisture can be determined by different methods [17]: (i) the gravimetric method; (ii) neutron scattering method; (iii) electrical resistance method; and (iv) tensiometer.

3.4.1   THE GRAVIMETRIC METHOD

After the soil has been irrigated, the soil is allowed to reach the equilibrium condition. After complete drainage, the soil samples from various soil depths are collected; the wet weight is taken and is placed in an oven at 105 °C for 48 h. The soil is then reweighed and water content is determined on dry basis.

$$Pw = 100[(Sw - Sd)/Sd] \tag{2}$$

where: Pw = soil moisture content on dry basis, %; Sw = the weight of wet soil; and Sd= the weight of dry soil. The volumetric moisture percentage can be calculated by multiplying the gravimetric moisture percentage by the apparent density of the soil. The soil moisture characteristics for different types of soils are shown in Table 3.3.

TABLE 3.3.   Soil Moisture Characteristics

Soil type	Apparent density	Gravimetric moisture percentage, Eq. (2)	
	g/cm3	Field capacity	Wilting point
		%	%
Clay	1.20	40	30
Clay loam	1.80	30	22
Loam	1.40	18	8
Loamy sand	1.55	8	4
Sand	1.60	5	2
Sandy loam	1.50	14	5

Available soil moisture = Field capacity – permanent wilting point.

3.4.2   NEUTRON SCATTERING METHOD

Method of soil moisture estimation by Neutron and Gama ray scattering was first advocated by Belcher and Gnykendall [7]. Holmes and Turner [7] developed a portable apparatus for field use. It consists of a radium beryllium neutron source, a slow neutron, boron trifluoride detector with preamplifier incorporated in a cylinder called the "probe" and a scaler for counting the slow neutrons. The whole equipment can operate with 6-volts DC. In this device, the alpha particles originating from radium source hit the beryllium atom and give rise to fast moving neutrons, which are projected, into the soil. Hydrogen atoms present in the soil water slow down these neutrons. Higher is the soil moisture content, quicker is collision with the hydrogen atoms, with further slowing of neutrons. A measure of these slow neutrons is taken as a measure of amount of water present in the soil. The instrument is calibrated so as to convert the readings into the soil moisture status [7].

3.4.3 ELECTRICAL RESISTANCE METHOD

Bouyoucos [7] developed porous gypsum blocks with electrodes to measure electrical conductance that indirectly indicates the soil moisture status (Fig. 3.9). This electrical resistance changes with change in soil water content. The water content of the block will come in equilibrium with the soil water content. The resistance readings for a given soil and water content are influenced by salinity. The increase in salinity reduces the resistance. It is found to be not suitable for sandy soils.

3.4.4 TENSIOMETER

The tensiometer measures range of soil moisture content within which the plant roots actively absorb water. The main components of the equipment are porous ceramic cup attached to the lower end of the tensiometer and a vacuum gage. The cup and the rigid tubes are filled with dye water. The water in the cup is in hydraulic contact with soil water in the pores. When the soil dries, it sucks water from the cup through the pores. As a result, tension develops within the system, which gradually increases as the soil gets dried. The suction or tension is measured through the vacuum gage. If the soil is wet (at field capacity), there is no tension and the gage indicates "zero" reading. These tensiometer measurements are useful to find energy status of the soil and water. The tensiometer cannot be operated under soil moisture conditions with tension greater than one bar (Fig. 3.10).

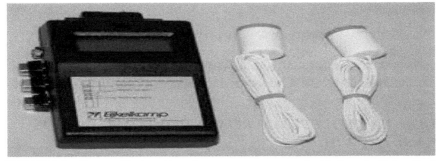

FIGURE 3.9   Gypsum blocks with an instrument to measure electrical conductivity of soil moisture.

FIGURE 3.10   Tensiometer.

3.4.5   IRRIGATION REQUIREMENT

The consumptive use (CU) of any crop is the total amount of water that must be replenished in the soil to bring it to its complete maturity. It includes the water lost by transpiration and evaporation. The climate, temperature, and the relative humidity affect the evapotranspiration. The transpiration rate also varies depending upon the degree of plant development, the amount of foliage and the nature of leaf surface. The ratio between evapotranspiration (ET) and evaporation during the growing season is not a straight line but changes with the stage of plant growth. In the beginning the ratio of ET to Epan is low. The water deficit can be calculated by using the following equation:

$$\text{ASMD} = [(\text{FC} - \text{Aw})/\text{FC}] \times 100 \qquad\qquad (3)$$

where: ASMD = available soil moisture depletion in %; Aw = available soil moisture, fraction; and FC = field capacity in fraction.

3.4.6   THE INFLUENCE OF RETARDED TRANSPIRATION ON PLANT GROWTH

In arid regions, the plant growth depends upon the texture and structure of soil, and also the tolerance of plants to water with the high salt content. Transpiration per unit leaf area was increased by 80% for all salinity levels as the relative humidity decreased from 90% to 25% [4, 5].

3.4.7   FERTILIZATION THROUGH DRIP IRRIGATION SYSTEM

All soluble fertilizers can be fertigated through the drip system. Chemicals that are only partially soluble should not be fertigated. The fertilizers can be classified into three groups depending upon their effects on the pH of the soil (acidic, alkaline and neutral). The salt index is defined as the degree of increase in the osmotic pressure of the soil solution due to the fertilizer. The constant use of urea and ammonia salts acidifies the soil. By this the capacity of the plant to absorb water is reduced. The chloride increases the osmotic pressure in the plant. Excess use of sulfates will develop specific toxic effects on the plants. Plant damage resulting from high concentration of salts can be detected by noting root injury, burning of leaf tips etc. In drip irrigation, it is advisable to use the fertilizer in split doses.

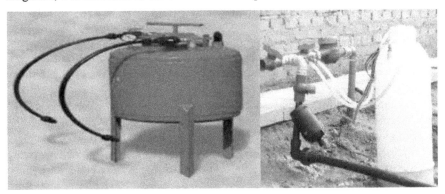

FIGURE 3.11   Fertilizer tank for drip irrigation.

The fertilizer tanks can be used for mixing the fertilizer. The tank can be connected to the irrigation pipe by creating a pressure differential between the tubes entering and leaving the tank (Fig. 3.11). With the fertilizer tank, the dilution ratio is not constant and changes with time. It is very low at the beginning and goes on increasing. The operation starts with prior dissolution.

3.4.8   IRRIGATION REQUIREMENT AND CROP YIELD

The crop yield under drip-, furrow-, and sprinkler-irrigation is shown in Table 3.4. In Israel, major crops under trickle irrigation are: apple, pear, grape, peaches, apricot, plums, almonds, oranges, grape fruit, lemon, banana, guava, mango, tomato, green pepper, eggplant, melon, cucumber, cotton, sugarcane, sorghum, sweet corn, etc.

TABLE 3.4 Comparison of Crop Yield with Different Irrigation Methods and Crop Water Use, in Israel

Crop	Water applied, inches	Crop yield, tons/acre		
		Drip	Furrow	Sprinkler
Cucumber	26.4	19.6	9.6	—
Musk melon	25.8	17.2	—	9.6
Peppers	55.8	3.8	—	1.9
Sweet corn	26.6	4.9	—	2.1
Tomato	38.7	26.0	—	15.6

Mature and bearing trees were adopted to trickle after the crops under conventional irrigation method did not suffer and adopted readily to new water regime. The reduction of wetted area did not cause severe harm to roots in the summer in the arid regions where subsequent rainfall or other winter applications of water for leaching purposes assumed resumption of root renewal. In Israel, filtration system for drip irrigation consists of vortex filters for effective filtration of water containing appreciable quantities of sand. Chemical clogging of emitters due to water and fertilizers deposits continues pose challenge. Clogging due to hard water deposits is limited only to certain areas and conditions.

The Table 3.5 shows plant response to the drip method, for several crops. The data indicate: Larger and heavier fruit trees; higher yield; fuller and earlier color pigmentation; higher sugar content or fat percentage; and other high grade qualities. Recent research studies aim to minimize severe damage to some fruit bearing trees, because of dry weather. Micro climatic control processes at peak hours of dryness have already shown some progress. Studies conducted in research plots and commercial fields give valuable knowledge on effectiveness in water consumption, and crop yields.

Water saving in vegetable crops reached mainly by avoiding watering beyond the field capacity compared to sprinkler irrigation and by short intervals. Soil moisture tests show clearly that low water tension is directly proportional to the time span between intervals. A stage can be reached in trickle irrigation, where saturation develops rapidly when daily rates are based on the optimal average rates in common sprinkler practices. This leads to the conclusion that reduction of irrigation coefficients should be evaluated for drip irrigation.

In South Africa, Blass type drip irrigation (Agriplas Agricultural Plastics) was commercially used for tomato, table grapes, apples, guava etc.

In California, research on with strawberries showed that drip system was able to save 50% of water consumption for conventional irrigation, and gave equal yield and better quality and size grading of fruit. The salt content was decreased by 400% under drip system compared to that under furrow irrigation, while no tip burn occurred under drip system. It is also seen that number of plants per ha can be increased by 50% with equal performance of plants [16].

TABLE 3.5   Crop Yield Under Three Irrigation Systems

Crop	Drip		Furrow		Sprinkler	
	Id	**Yield**	**Id**	**Yield**	**Id**	**Yield**
	mm	**Tons/ha**	**mm**	**Tons/ha**	**mm**	**Tons/ha**
			Israel			
Apples	900/year	120	–	–	900/year	100
Cotton	550	5.7	–	–	550	4.7
Table grapes	420	289	–	–	–	–
Green pepper	—	10.5	–	–	–	7.5
			California, USA			
Tomato	–	75.8	–	61.3	–	–

Apples: Under semiarid to semihumid zones, drip irrigation gave 20% higher yield compared to sprinkler method.

Cotton: Under semiarid zones, irrigation treatments were based on 0.6, 0.7, and 0.9 Epan evaporation.

Table grapes: In sandy clay soils, irrigation was based on Epan coefficients.

Green pepper: Under arid to semiarid conditions and in sandy clay soils: vegetative growth was faster and superior in drip irrigation. Early maturity was observed.

Tomato: In sandy clay soils: vegetative growth was faster and superior in drip irrigation.

Id = Total irrigation depth, during the crop growing season or per year (for apples).

In some orchards in South Australia, effluent water of high salinity has been successfully used in drip irrigated wine grapes. The effluent water was filtered with a mesh filter at the pumping station and a secondary mesh filter adjacent to the filters. Algae growth in the pipeline was controlled by periodic slug dosing with Calcium hypochlorite.

Polythene plastic mulch has effectively complemented drip irrigation on vines, strawberries, citrus, grapes, peaches, apples and rock melons. Its use has improved weed control, water saving and crop yields.

In deep sandy soils, drip method is suitable for crops like citrus, provided excessive leaching of nutrients does not occur. Emitters must be placed around each tree so that 60% of soil volume is wetted.

TABLE 3.6   Effects of Irrigation Treatments on Grape Vines in NSW, Australia

Treatments	Evaporative replacement	Water depth	Yield 1970 1971		No. of branches (wines)	Weight of bunches	Weight of berry
		inches	Tons	Tons	1971	lbs	lbs
Drip irrigation							
Daily, all season	0.50	15.6	12.7	10.8	180	109	1.64
Once in two days, all season	0.50	13.1	11.0	10.3	187	99	1.55
Daily, after veraison	0.50	14.2	9.0	7.6	180	75	1.08
Daily, to veraison	0.25	8.3	9.0	8.9	175	–	1.55
Daily, all season	0.25	10.0	10.0	9.3	178	91	1.55
Furrow irrigation							
Irregular, only under severe moisture stress		9.9	7.5	6.1	210	51	0.89
Irregular, to harvest after 3.5 evaporation		12.1	12.1	11.4	195	105	1.55
Irregular, all season after 3.5 evaporation		21.4	12.8	11.4	185	111	1.57

Veraison is a viticulture (grape-growing) term meaning "the onset of ripening." It is originally French, but has been adopted into English use.

In New South Wales (Australia), comparative studies of plant water stress in different irrigation treatments were conducted in grape vines (Table 3.6). Trickle irrigation treatments were on daily and on alternate day at two levels of evaporation replacement during different times of the growing season. Furrow irrigation treatments included irrigation up to harvest only and up to post harvest, and in both cases after 9 cm of accumulated class "A" pan evaporation. Each treatment was replicated four times and average value was calculated. The weight of berry was higher for drip irrigation compared to the furrow irrigation, but the differences among the drip irrigation treatments were nonsignificant.

Another trial was conducted to evaluate trickle irrigation technology for the establishment of apple, citrus, peach and vine plantings. Four irrigation treatments based on replacement of 40, 60, 70 and 80% of the evaporation from a wetted soil proved adequate for apples, citrus, vines. Black polyethylene plastic mulch was used in conjunction with these treatments. The tree growth was not significantly different among all irrigation treatments. Trees with the black mulch gave better growth than those with no plastic mulch.

Fertigation in drip system has indicated that only urea and other soluble fertilizers can be successfully applied. In many cases trace element deficiencies occur not because of a total lack of the particular element but because of nonavailability as a result of unavoidable soil reactions. The application of water-soluble phosphates through drip irrigation is risky since the water contains Calcium or Magnesium, which will combine with Phosphate and will cause clogging due to chemicals that are deposited in the emitters.

In Israel, drip irrigation trials were conducted on various vegetable crops. The onion size was 6.15–7.20 cm in conventional irrigation compared to 7.71 cm in drip irrigation. Similarly, the weight of ladyfinger was 5.47–6.35 kg per plant in surface irrigation compared to 6.88 kg/plant in drip irrigation.

At the Central Arid Zone Research Institute (ICAR), Jodhpur– India, research studies have shown that drip irrigation is more suited to high value widely spaced vegetables and plantation crops in the sandy arid plains. Table 3.7 presents effects of irrigation methods on yield, WUE and net returns of bottle guard (*Loki*, local name in India) and ridge guard. The yield of drip irrigated bottle guard was much higher compared to those in sprinkler and furrow systems. The differences in yield among sprinkler and furrows were not significant. In the case of ridge guard, the differences in crop yield were not significant. The net profit per unit area per day for the two crops was high in drip irrigation system than the values for other two irrigation systems. Plant root distribution showed that the roots were within 30 cm of soil depth, in the sprinkler system. In drip irrigation system, roots were distributed up to 60 cm soil depth thereby permitting a better exploitation of soil by the roots.

TABLE 3.7 Crop Yield, Water Use Efficiency (WUE), and Net Returns: Bottle Guard (Loki) and Ridge Guard Under Three Irrigation Systems

Irrigation method	Growth period	Total Water depth	Yield in tons	WUE	Net return, Rs.	
	days	cm	per ha	per ha per unit of water	per ha	per day of crop season
Bottle guard (Loki)						
Drip	133	74	558	7.54	20,656	155
Sprinkler	133	84	386	7.59	13,527	198
Furrow	133	84	380	4.52	13,099	99
Ridge guard						
Drip	123	74	120	1.63	4,207	34
Sprinkler	123	84	100	1.19	2,975	24
Furrow	123	84	107	1.28	3,197	26

3.4.9 EFFECTS OF IRRIGATION TREATMENTS ON WUE AND CROP YIELD IN COIMBATORE – INDIA

The Table 3.8 shows WUE for various crops at TNAU Research Farm, Coimbatore – India. Perforated alkathene pipes with sockets were used for the drip irrigation trial plots. Tomato yield was 9000 kg/ha in drip irrigation compared to 7000 kg/ha in the control. The water saving under drip irrigation varied from 60 to 80%, among all crops. From the results in Table 3.8, it is concluded that drip irrigation saves water quantity and more yield is obtained. Also 0.25 to 0.33 of water used in control was sufficient for plant growth, in all drip-irrigated crops.

TABLE 3.8 Water Use and Yield of Various Crops in Drip and Surface Method of Irrigation at TNAU, Coimbatore – India

Crop	Water use, cm		Yield, kg/ha		Rainfall, cm
	Drip	Control	Drip	Control	
Banana	58.0	243.0	15.3 kg/plant	16 kg/plant	34.0
Beetroot	17.7	85.8	887	571	...
Bhendai, okra	8.8	53.5	11,310	10,000	24.2
Brinjal, eggplant	24.5	69.2	12,300	12,400	17.2
Chilies	41.8	109.7	6080	4233	20.8
Radish	10.8	46.4	1186	1045	...
Sweet Potato	25.3	63.1	5888	4244	12.1
Tomato	12.1	50.2	9000	7000	22.1

3.4.9.1 COTTON CBS 156

A replicated trial for studying effects of types of emitters on the plant growth of cotton (var. CBS 156) was evaluated on a plot size of 8×8 m^2 for following treatments:

- T1 – Drip method using 12 mm diameter alkathene pipes with 'hole and socket' emitter.
- T2 – Drip method using 12 mm diameter alkathene pipe with microtube of one mm diameter.
- T3 – Polyethylene hose stitched in the form of tubes having 2.5 cm width.
- T4 – Control plot using conventional furrow method.

All these four treatments were replicated 6 times. A typical drip irrigated cotton field is shown in Fig. 3.12. The PE hose was laid on the ground in a straight line and the ends were tied to a pole. A small depression of 2 cm was made on

FIGURE 3.12   Drip irrigation in cotton.

the ground and the tube was placed along it for the whole length of 8 m. This prevented the hose movement from its place due to wind movement. The water was oozing out along the stitches wetting a continuous band along the row. For the microtube plots, the length of microtubes was adjusted for uniform flow. In the drip method, using alkathene tube with holes, the diameter of the holes was adjusted for uniformity of flow. For the control plots, the furrows were made at 75 cm center to center. The irrigation was applied based on the pan evaporation coefficient for the experimental site. The control plot was irrigated at 50% ASMD and the quantity was recorded. The CBS 156 cotton seeds were sown in the plots and were irrigated daily.

The data for plant growth, yield, water depth are given Table 3.9. Data reveals that the plant growth was exceedingly well in drip and micro tube plots compared to control and PE stitched hose plots. The yield was maximum in the drip plots followed by micro tube and control plots. The yield was lowest in the plots with PE hose. The yield reduction may be due to lower percentage of the wetted soil volume in these plots.

TABLE 3.9   Effects of Four Irrigations Treatments on Cotton Yield at TNAU Farm

Treatment	Total water depth	Yield	Rain-fall	Avg. wetted area
	cm	Kg/ha	cm	%
T1, Drip with hole and socket	13	3255	13	43
T2, Drip with micro tube	13	2864	13	40
T3, Drip with the PE hose 2.5 cm wide	26	2213	13	30
T4, control plot with furrows.	70	2604	13	—

3.4.9.2 BANANA

The Robusta banana was planted to evaluate the crop response to drip irrigation and control method (Fig. 3.13). Main pipes were 50 mm and laterals of 12 mm diameter. The plant-to-plant spacing was 1.8 m. The laterals were placed around the plant in treatment T1, and two laterals were put on either side in treatment T2. The inlet pressure at lateral was 2 psi. Assuming daily evaporation of 10 mm, and that each hole wets an area with diameter of 60 cm around the plant, the water requirement was calculated as 2.8 L/plant/day. After the 7th month at flowering, the water quantity was doubled. Water depth was recorded with water – meter. In control method T3, the irrigation was given at 50% ASMD. Tables 3.10 and 3.11 show that yield differences were not significant among all treatments. The yield was found on par with the usual conventional method.

FIGURE 3.13   Drip irrigation in banana.

TABLE 3.10   Effects of Drip Irrigation on Banana Yield

Total water depth, cm		Avg. weight of bunches, kg		Rainfall, cm
Drip	**control**	**Drip**	**Control**	
50	200	20	22	40

TABLE 3.11   Weed Growth and Flowering in Banana Plantation

Treatment	Mean weight of weeds	Percentage of flowered plants after	
	Kg	**9 months**	**10 months**
Drip	5	30	55
Control	15	14	35

It can be observed that even though yield was reduced by 2 kg/plant, the water saving was 0.75 of the control system. It was also noticed that plants in the plot irrigated by drip method flowered earlier than those in control. Statistical analysis of data showed no significant differences in yield among all treatments. Drip plots only used 25% of the water applied in the control plots.

FIGURE 3.14 Drip irrigated papaya field.

3.4.9.3 PAPAYA

Observational trials were conducted for papaya at research farm of TNAU (Fig. 3.14). The treatments included drip irrigation and control method. The yield in drip plot was increased by 69% compared to the control plots. Drip irrigation used 73.4 cm of total water compared to 228.5 cm for the control method, excluding rainfall in both methods (Table 3.12).

TABLE 3.12 Effects of Irrigation Methods on Papaya Yield

Method	Water depth	Yield per plant	Rainfall
	cm	kg	cm
Drip	73.4	23.5	82
Control	228.5	14	82

3.5 WATER MOVEMENT AND DISTRIBUTION

The water movement in a soil affects the extent of soil wetted volume/depth and concentration of salts in the root zone. In drip irrigation, this depends on the factors, such as: soil liquid limit, soil plastic limit, soil porosity, soil hydraulic conductivity, soil water status, rate of infiltration into the soil, emitter discharge, initial soil moisture content before application, water table level, irrigation duration, emitter spacing, soil evaporation, and root suction. The water distribution pattern can be determined by direct observation by excavations, gravitational moisture analysis, tensiometer readings, use of gypsum blocks, Neutron scattering method.

There are three phases for a water distribution in the root zone under single point water source: A transmission zone, a wetting zone, and a wetting front. In transmission zone, the soil becomes saturated. In the wetting zone the water flows into the soil in the direction of minimum resistance and medium gradient. The moisture content decreases proportionally to the distance from the point source. The moisture in the extreme boundaries of the wetting zone balances and equals the original soil moisture. All the three phases are continuous and a definite boundary cannot be marked. The shape slope and depth of the wetted soil is a function of the flow rate of an emitter, the soil type and the rate of evapotranspiration. In a given time, higher the discharge, deeper the penetration and wider the lateral flow.

Advance of the wetting front is also affected by soil moisture characteristic curve, emitter flow rate and spacing. A typical water distribution under a point source is shown in Figs. 3.15 and 3.16.

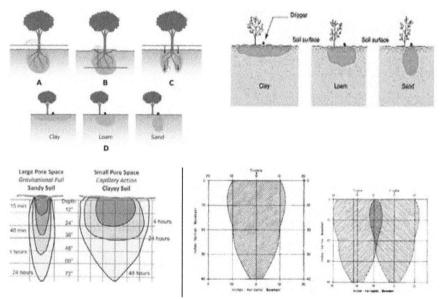

FIGURE 3.15   Water distribution under a point source and for heavy and sandy soils.

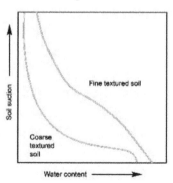

FIGURE 3.16   A typical soil moisture characteristic curve.

To study the soil moisture distribution, replicated trials with different application rates under drip irrigation were conducted at the research farm of TNAU, Coimbatore. The soil at the site is clay loam. The treatments were:

T1 Water delivered at 30 lph for 20 min/day.
T2 Water delivered at 20 lph for 30 min/day
T3 Water delivered at 10 lph for 60 min/day.
T4 Water delivered at 5 L/hr. 120 min/day.

To eliminate the effects of moisture removal by roots, the experiments were conducted on the field free of weeds. Drippers were spaced well apart to prevent possible overlapping of wetted areas between emitters. The four treatments were replicated 5 times. The experiment was continued for six weeks and after this trenches were dug through the central line of wetted pattern. From this, the shape and size of wetted pattern front were observed. The boundaries of wetted areas were clearly visible, however the wetting depth it was not clearly marked. The soil moisture content was determined by taking soil samples at different locations and depths around the dripper. Table 3.13 indicates the mechanical analysis of soil at the site.

TABLE 3.13   Soil Texture at the Experimental Site

Soil depth	Percentage distribution, %		
	Coarse sand	Fine sand	Silt clay
cm	%	%	%
50	7	20	73
50 to 100	10	18	74
100 to 150	10	20	70
Average value of samples at different locations.			

It was observed that with lower emitter flow rate for longer time, the depth of wetting was higher than the high flow rate for shorter time. It shows that the water distribution is directly dependent on both the discharge rate of dripping and duration of irrigation. The rapid rate of soil water increases during the initial period of application in all the treatments causing saturation below the dripper point. It spreads to a wider area on the ground surface for a high rate of application. However, the horizontal movement of water was more spread than the vertical movement for the same quantity of water applied in all treatments. Therefore, the lower rate of application for longer duration is more suited to keep the soil moist and for efficient plant growth.

3.5.1   EFFECTS OF DISCHARGE RATE OF AN EMITTER

The discharge rate of a dripper usually exceeds the infiltration capacity of the soil. The diameter of the wetted surface depends on the rate of flow of emitter and the infiltration of the soil.

$$D = 2 \times r = \{[q \div I] \times [4/\pi]\}^{0.5} \tag{4}$$

where: D = diameter of the wetted surface, cm; r = radius of the wetted surface, cm; I = Infiltration capacity of soil, cm/min; and q = discharge rate, cm^3/min.

During irrigation, the infiltration rate changes according to the changes in soil structure and changes in forces, which cause the water to move into soil. The water enters through the soil aggregates initially and then it passes through the fine capillary pores [7]. In clayey soils, there is swelling of clay minerals. This reduces hydraulic conductivity at saturation. From experiments, it was noticed that the area occupied by the total wetted soil is greater for high discharge. An empirical equation for describing water spreading has been described by Goldberg [4, 5].

$$D = a\,Q^{0.5} + b\,qQ^{0.5} + C = C + [a + bq]Q^{0.5} \tag{5}$$

where: D = diameter of spread, cm; q = discharge, lph; Q = water applied in liters; and the regression coefficients are a, b, c that must be found using the data from field studies. For sandy soils, Goldberg and others have revealed that the yield was much higher for drippers with high daily discharge rate [4–6, 17, 18].

In drip irrigation, the plants are more sensitive to temporary and irregular dry periods. A comparison of wetting pattern under the drippers was made for citrus in a sandy soil by Cole in Australia [1, 2]. He found that with one lateral line of drippers, the mid row was not wet and excess water was found passing beyond the root zone. Using two laterals of drippers, most of the soil was wetted, with only the soil surface between the drip lines remaining dry. However, with drag hoses, all the soil was wet. The wetted soil volume from a single line of drippers is much less than the volume wetted by two lines of drippers. Since the same quantity of water is applied in each case, leaching must be greater with the single line of drippers.

3.6   LAYOUTS OF DRIP IRRIGATION SYSTEM FOR DIFFERENT CROPS

The field layout for a particular crop depends on the crop types, field and hydraulic situations. The lateral spacing and the selection of dripper size must be based on crop characteristics, soil properties, water quality and the agro-technical practices. The drip irrigation layouts can be classified into three main categories: For field crops; for closely spaced vegetables; and for orchard crops.

For field crops, such as sugarcane, the emitters are fitted on the laterals at close spacing (Figs. 3.16 and 3.17). The sugarcane is a shallow rooted crop with most of its roots in the upper soil layer of 0–30 cm soil depth. Sugarcane crop is sensitive to changes in soil water tension. There has been 100 to 200% increase in yield with water saving of 25 to 33%, under drip irrigation. Drip irrigation system for corn is shown in Fig. 3.18.

FIGURE 3.16   Drip irrigation in sugarcane: Left, surface drip irrigation; Right, Subsurface drip irrigation.

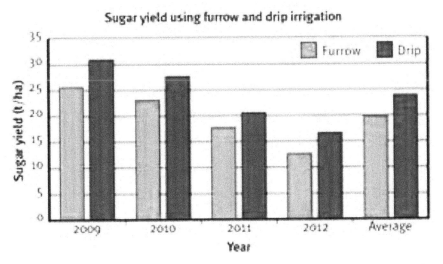

FIGURE 3.17   Sugarcane yield under furrow (Left charts for each year) and drip irrigation (Right charts for each year): Australia.

FIG. 3.18   Drip irrigation layout for corn.

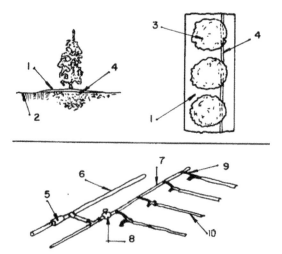

1. Organic matter mulch.
2. Soil surface.
3. Plant.
4. Emitter integrated on the line or in-line drippers.
5. Filter.

6. Principal or main line.
7. Secondary line or sub main.
8. Optional: Gate valve.
9. Connecting tube (or spaghetti).
10. Lateral line.

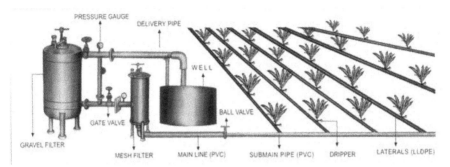

FIGURE 3.19 Micro irrigation system in vegetable crops.

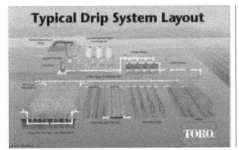

Toro irrigation, California.   Bucket type gravity drip irrigation.

FIGURE 3.20   Examples of drip irrigated vegetable crops.

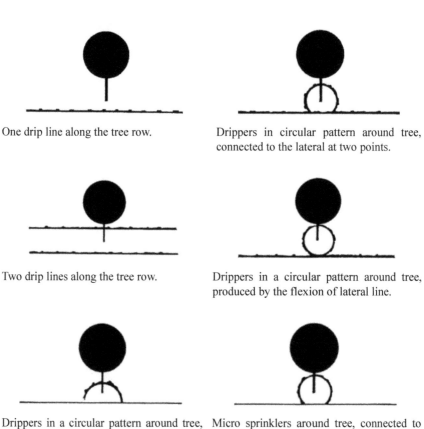

One drip line along the tree row.

Drippers in circular pattern around tree, connected to the lateral at two points.

Two drip lines along the tree row.

Drippers in a circular pattern around tree, produced by the flexion of lateral line.

Drippers in a circular pattern around tree, connected to the lateral at one point.

Micro sprinklers around tree, connected to the lateral line at points.

FIGURE 3.21   Different arrangements of micro irrigation laterals for fruit orchards [7].

Drip irrigation in citrus (top) and apple (bottom).   Drip irrigated mango.

FIGURE 3.22   Selected examples of drip irrigation in fruit orchards.

Closely spaced crops: All vegetables (eggplant, okra, beans, onion, tomato, peppers, peas, cauliflower, cabbage, and etc.) are closely spaced crops (Figs. 3.19 and 3.20). The plants at maturity cover the entire soil surface. These crops respond favorably under low soil water tension. The cost per acre of the drip system depends mainly on the lateral lines and is about Rs. 3000 per acre using low cost alkathene pipes. For okra, peppers and radish, drip system has been found to be much beneficial by increasing yield and water saving. For areas of water scarcity such as Tamil Nadu – India, farmers largely adopt dry land cultivation practices. The drip irrigation system helps the farmer to judiciously use his available land and water.

For cotton, also the drip method is very effective. The yield of cotton CBS 156 was increased by 10% compared to the control method. The water saving was nearly 60%. Due to a cash crop, the farmer gets more profit with cotton compared to vegetables or other crops. Moisture during the flowering and prematuring stages is more critical. With excessive irrigation, production may be reduced. In drip irrigation, one can fully control the timing as well as quantity of water applied.

Orchard crops are particularly suited to drip irrigation. Banana, grapes, citrus, mango, guava, avocados, ber (Indian fruit: *Ziziphusmauritiana*), papaya, etc. are cultivated by drip irrigation in various parts of the world. In Israel, drip irrigation has resulted in higher yield of olives, data palms, almonds, walnut, etc. Different arrangements of drip irrigation layout are shown in Figs. 3.21 and 3.22.

Banana is a tropical plant requiring large quantities of water. The normal spacing is 1.8×1.8 m². The soil must be kept at low soil water tension throughout the growth period. The limited wetted soil results in a restricted root zone volume but the yield is not affected. The extent of anchoring must be considered in areas of high winds and intensity. Without anchoring, it may result in overturning of the tree if the plant does not have its roots all around due to the restriction of the wetted zone. This can be minimized if three or four emitters of low discharge are installed around each banana plant (Fig. 3.23). The plants, which do not receive optimum water, are not developed properly. The 90% of the total roots are found in the upper 0 to 30 cm of soil depth. High discharge emitters cause the active roots to spread laterally and to greater depths.

Grapes respond favorably to drip irrigation. During berry formation, it is essential to maintain a high level of soil moisture, which depends on grape variety and cultural practices (Fig. 3.23).

Papaya is very sensitive to soil water tension (Fig. 3.23). The normal plant spacing is 1.8×1.8 m². Single lateral per each row is sufficient for most soil types. Research studies, in drip irrigated fruit trees at TNAU, have shown water saving of about 60% and the increase in yield by 69%.

Drip irrigation has been found beneficial in other fruit trees such as: apple, apricot, plums, etc. In Israel, olive, date palms etc. are grown by drip irrigation and the yield had been found to be higher than those, which are irrigated by other methods.

Drip irrigation in banana.

Drip irrigation in grape vines.

Drip irrigation in papaya.

Drip irrigation in mulberry.

FIGURE 3.23 Drip irrigation layouts for: Banana (Left top); Papaya (Left bottom); Grape vines (Right top); and mulberry (Right bottom).

3.7 ECONOMICS OF DRIP IRRIGATION SYSTEM [11 TO 15 AND 23, 24]

The high initial cost of trickle irrigation system is of concern among farmers, who can get government subsidy. The high cost is attributed to the costs of PE lateral tubes, emitters, accessories and valves, chemigation and filtration equipments, and permanent other equipments, etc. A more comprehensive approach is an analytical evaluation of this high cost with:

- total expenditure per unit area of a crop.
- the alternative costs for different irrigation methods under similar conditions.
- the comparative potential rate of returns.
- the specific significance of alternative components in the methods compared, e.g., labor against equipment.
- the relative importance of individual components to overall design.
- the possibility of drawing the optimal design within the framework of the method chosen.

Research studies in Israel have contributed to offer a systematic analysis for the selection of adequate design of a trickle irrigation system for a particular crop. The economic analysis (cost per unit area) included the cost of laterals, mains, emitters and the header. It included necessary factors of design engineering significance. It has been observed that: Lateral spacing greatly influences the cost compared to the emitter spacing; emitter spacing influences the cost more than the emitter discharge. Thus for a given quantity of water, the optimal design will tend towards maximum distance and spacing.

The emitter discharge is selected based on water use by the crop, crop practices, crop type, soil moisture capacity, soil texture and structure, and emitter types and characteristics. Length of laterals and mains/submains depends only on the size of area and crop spacing. These lengths are not influenced by the system operating pressure, and the emitter characteristics, spacing and discharge. Diameters of the lateral, main and submain lines are selected according to the frictional losses, operating pressure and system discharge. Design procedure is discussed in section 5.8 in this chapter.

The design of pipe layout must be selected for a minimum cost of the layout. There are numerous arrangements of pipe sizes that can be designed to meet the hydraulic situation of a given layout. Since the number of laterals in widely spaced crops (fruit trees) is less compared to vegetables and field crops, therefore drip system for orchards is of permanent type; is generally economical; and is easily adaptable in most of the world.

The initial cost of a sprinkler irrigation installation is generally major disadvantage. However, this cost is often not significantly higher than the costs of well-planned gravity irrigation systems when all land forming, conveyance, distribution, drainage and maintenance facilities are taken into account. The overall annual costs for sprinkler irrigation are often equal to and only a little higher than for surface irrigation systems, if full evaluation is made of the saving in water, the extra area that can be irrigated with the water saved, crop yields, quality and soil management aspects. The cost of a sprinkler irrigation system is different for areas of different sizes, and local market conditions. In India, investment cost for the basic portable sprinkler irrigation system will generally range from Rs. 8000 to 9000 per ha, whereas surface irrigation system of high-performance will cost about Rs. 3000 per ha depending on well depth, overhead tanks, land topography, etc. It is suggested that with all the additional benefits obtainable under sprinkler irrigation, the adoption of this method as the predominant irrigation method is well justified.

With drip irrigation, the situation is completely different. The cost is closely proportional to the number and length of distribution lines per ha. The price break-up and total sums per hectare are presented in this section grapes, orchard, banana orchard, coffee plantation and vegetable plots.

3.7.1 COST PER HECTARE: GRAPES AND BANANA ORCHARDS UNDER INDIAN CONDITIONS

Plant density: 3000/ha

Pipes: Alkathene or low density PE pipes

Costs (in Rupees):

9600 m of laterals 12 mm dia	4800.00
300 m of mains and submains 50 mm dia	1200.00
Other fittings and accessories, etc.:	200.00
Total cost (Rs./ha):	6200.00

3.7.2 COST PER HECTARE: VEGETABLE CROPS

13200 m of laterals 12 mm dia	6100.00
300 m of mains and sub mains 50 mm dia	1200.00
Other fittings and accessories, etc.:	250.00
Total cost (Rs./ha):	**7550.00**

For coffee and tea plantations, the drip irrigation is adopted to supplement the natural rainfall during the dry periods and under adverse climate. The laterals are laid on hill side slopes. In such situations, the submains are laid down slope to reduce frictional losses. The laterals are laid on the sides, across the slope for better uniformity. The system can be permanent or portable type so that the initial investment can be reduced considerably and may be used as need arises for a larger area. The economics of drip irrigation for 2 hectares is presented in the next section.

3.7.3   ECONOMICS OF DRIP IRRIGATION

The cost of drip irrigation is compared to the gravity irrigation system. It is assumed: Irrigation water is available at the location; the capital cost of well and pump-set is constant for the whole area, irrespective of method of irrigation; efficiency of drip irrigation is three times compared to that of gravity irrigation system; the overhead expenditure for well and pump is assumed the same for both irrigation methods; and cost of well, pump and land are not taken into account, since these are common for both irrigation systems. The economics of the drip irrigation is presented in this section for the following data:

Average cost of drip equipments, Rs./ha	6750
Average life of equipment, years	5
Total area of farm, ha	2
Total water available, ha-cm in a year	100 to 120
Rate of interest on capital investment,	% 12

Because the water resource is limited (only 100–120 ha-cm), the farmer can cultivate banana or sugarcane, which these are economically more profitable crops under drip irrigation. In the case of conventional irrigation, the farmer can cultivate cotton and other vegetable crops that require less water. During rainy season, he may raise dry crops only. Assuming that in both cases all the lands are cultivated with the limited water supply, the economics can be worked out as shown in Tables 3.14 and 3.15.

Table 3.14 shows the expenses under drip irrigation and the income from the crop for five successive years. The initial investment for pipelines for two hectares is assumed as Rs. 13,500. The average net gain per ha in using drip irrigation is Rs. 4,711, after deducting depreciation and interest for capital expenditure. In conventional irrigation the net gain is Rs. 3,290 as shown in Table 3.15. Data in Tables 3.14 and 3.15 indicate that the drip system saves water and increases crop yield compared to surface irrigation methods. The equipment designed and used at TNAU, Coimbatore is simple and economical, and it can be adopted by small and marginal farmers. In Tables 3.14 and 3.15, the economics of the system determines taking into account the depreciation and interest on capital investment, the net increase in income with different cropping patterns will be about Rs. 1,421 per year per hectare. Since the water quantity is limited, farmer can irrigate nearly three times the area under drip irrigation compared to gravity irrigation. The income of the farmers and the employment opportunities can also be increased in the villages. Procedure presented in Tables 3.14 and 3.15 can be used to compare the costs for other locations in other countries.

TABLE 3.14 Cost (Rupees for 2 ha) of Using Drip Irrigation Based on Assumptions That Are Mentioned in this Section

Crop	Parameters								
	Culti-vation ex-penses	Depre-ciation	Inter-est	Total ex-penses	Wa-ter use	Crop yield	In-come	Net income	Re-marks
1	2	3	4	5	6	7	8	9	10
—	Rs.	Rs.	Rs.	Rs.	cm	Kg	Rs.	Rs.	—
First year									
Banana, 1ha	9000	–	–	9000	60	45,000	15,000	–	–
Cotton, 1 ha	6000	–	–	6000	15	2,500	11,000	–	Jul–Dec
Eggplant, 1 ha	3000	–	–	3000	25	15,000	6,000	–	Jan–May
Depre-ciation and interest on capital	—	2700	1620	4320	–	–	–	–	–
Total	**18,000**	–	–	**22,320**	–	–	**32,000**	**9,680**	–
Second year									
Banana, 1 ha	9000	–	–	9000	50	45,000	15,000	–	–
Cotton, 1 ha	6000	–	–	6000	25	2,500	11,000	–	–
Tomato, 1 ha	3000	–	–	3000	75	15,000	6,000	–	Jan–May
Depre-ciation and interest on capital	—	2700	1296	3996	–	–	–	–	–
Total	**18,000**	–	–	**21,996**	–	–	**32,000**	**10,004**	–
Third year									
Sugarcane, 1 ha	5,000	–	–	5,000	70	86,000	9,000	–	1 year
Eggplant, 1 ha	3,000	–	–	3,000	25	15,000	6,000	–	–
Cotton, 1 ha	6,000	–	–	6,000	15	2,500	11,000	–	–

Crop	Parameters								
	Culti- vation ex- penses	Depre- ciation	Inter- est	Total ex- penses	Wa- ter use	Crop yield	In- come	Net income	Re- marks
1	2	3	4	5	6	7	8	9	10
—	Rs.	Rs.	Rs.	Rs.	cm	Kg	Rs.	Rs.	—
Depre- ciation and interest on capital	—	2700	972	3672	–	–	–	–	–
Total	**14,000**	–	–	**17,672**	–	–	**26,000**	**8,328**	–
				Fourth year					
Sugarcane, 1 ha	5,000	–	–	5,000	70	86,000	9,000	–	–
Cotton, 1 ha	6,000	–	–	6,000	15	2,500	11,000	–	–
Chilies, 1 ha	3,528	–	–	3,528	40	15,000	6,000	–	–
Depre- ciation and interest on capital	—	2700	648	3348	–	–	–	–	–
Total	**14,528**	–	–	**17,876**	–	–	**26,000**	**8,124**	–
				Fifth year					
Banana, 1 ha	9000	–	–	9000	60	45,000	15,000	–	–
Cotton, 1 ha	6000	–	–	6000	15	2,500	11,000	–	Jul– Dec
Eggplant, 1 ha	3000	–	–	3000	25	15,000	6,000	–	Jan– May
Depre- ciation and interest on capital	—	2700	324	3024	–	–	–	–	–
Total	**18,000**	–	–	**21,024**	–	–	**32,000**	**10,976**	–
Grand total	–	–	—	–	–	–	—	**47,112**	—

Average net income, Rs. per year per ha = $[47122/(5 \times 2)]$ = Rs. 4,711

TABLE 3.15 Cost (Rupees for 2 ha) of Using Gravity (conventional) Irrigation Based on Assumptions That Are Mentioned in this Section

Crop	Parameters								
	Cultivation expenses	Depreciation	Interest	Total expenses	Water use	Crop yield	Income	Net income	Remarks
1	2	3	4	5	6	7	8	9	10
—	Rs.	Rs.	Rs.	Rs.	cm	kg	Rs.	Rs.	—
First year									
Sorghum, 1ha	620	–	–	620	Dry crop	52, grain 3 (T), S	1,200	–	Dry crop Jun–Dec
Cotton, 1 ha	6,000	–	–	6,000	60	2,400	10,000	–	Jun–Dec
Eggplant, 1 ha	3,000	–	–	3,000	70	14,000	5,600	–	Jan–May
Total	**9,620**	–	–	**9,620**	–	–	**16,800**	**7,180**	–
Second year									
Sorghum, 1ha	620	–	–	620	Dry crop	52, grain 3 (T), S	1,200	–	Dry crop Jun–Dec
Cotton, 1 ha	6,000	–	–	6,000	60	2,400	10,000	–	Jun–Dec
Tomato, 1 ha	3,000	–	–	3,000	60	14,000	5,600	–	Jan–May
Total	**9,620**	–	–	**9,620**	–	–	**16,800**	**7,180**	–
Third year									
Sorghum, 1ha	620	–	–	620	Dry crop	52, grain 3 (T), S	1,200	–	Dry crop Jun–Dec
Cotton, 1 ha	6,000	–	–	6,000	60	2,400	10,000	–	Jun–Dec
Okra, 1 ha	3,000	–	–	3,000	70	12,000	4,000	–	Jan–May
Total	**9,620**	–	–	**9,620**	–	–	**15,200**	**5,580**	–

Crop	Parameters								
	Cultiva-tion expenses	De-preci-ation	Inter-est	Total ex-penses	Wa-ter use	Crop yield	Income	Net income	Re-marks
1	2	3	4	5	6	7	8	9	10
—	Rs.	Rs.	Rs.	Rs.	cm	kg	Rs.	Rs.	—
				Fourth year					
Sor-ghum, 1ha	620	–	–	620	Dry crop	52, grain 3 (T), S	1,200	–	Dry crop Jun–Dec
Cotton, 1 ha	6,000	–	–	6,000	60	2,400	10,000	–	Jun–Dec
Egg-plant, 1 ha	3,000	–	–	3,000	65	14,000	5,600	–	Jan–May
Total	**9,620**	–	–	**9,620**	–	–	**16,800**	**7,180**	–
Fifth year									
Sor-ghum, 1ha	620	–	–	620	Dry crop	52, grain 3 (T), S	1,200	–	Dry crop Jun–Dec
Tomato, 1 ha	3000	–	–	3000	60	14,000	5,600	–	Jun–Dec
Egg-plant, 1 ha	3,000	–	–	3,000	70	14,000	5,600	–	Jan–May
Total	**6,620**	–	–	**6,620**	–	–	**12,400**	**5,780**	–
Grand total	–	–	—	–	–	–	—	**32,290**	**5 years**

A—Average net income, Rs. per year per ha = [32,290/(5 × 2)] = Rs. 3,290 for conventional irrigation
B—Average net income, Rs. per year per ha = Rs. 4,711 for drip irrigation
Net additional income by adopting to drip irrigation = A – B = 4711 – 3290 = Rs. 1421 per year per ha.
G = grain, S = Straw.

3.7.4 BENEFIT–COST RATIO, BCR [20 TO 29]

Micro irrigation is an accepted method of irrigation in India. Most of the farmers are convinced of the usefulness of the system, but the adoption is very slow due to its high initial investment cost. The cost of the drip system is about Rs. 15,000–25,000 per hectare for wide spaced crops like coconut, mango, pomegranate, oil palm. For MI system using 4 or 5 drippers per tree or one dripper at 50–75 cm spacing for closely spaced crops like vegetables, cotton, sugarcane, mulberry etc., the cost is about Rs. 50,000–60,000 per hectare.

The area under drip irrigation in India was only about 1000 ha in 1985 and increased to 60,000 ha in 1993. The present area under micro irrigation in the country is about 500,000 ha covering about 20 different crops like coconut, mango, oil palm, guava, sapota, pomegranate, *ber* (Indian jujube, *Zizyphus spp.*), lime, orange, grapes, cotton, sugarcane, vegetables (potato, tomato, onion, etc.), plantation crops (tea, coffee, cardamom, pepper, chilly) and flowers like rose, and mulberry (*Morus spp.*) etc. The projected area is estimated about 1 million hectare (1% of the irrigated area) in the year 2000–2005 AD and about 10 Mha (10% of the irrigated area) by 2020, though the crop area which is suited for the MI system is about 27 Mha as recorded by the government task force on the micro system. But the main constraint is its cost especially to the small and marginal farmers in India who comprise about 83% of total number of farmers and get only about 35% by income.

Benefit–cost ratio (BCR) was determined for various crops including sugarcane in Maharashtra State by contacting farmers (Table 3.16). Based on the study, it was found that the BCR for sugarcane with drip fertigation method was 2.66 compared to conventional (surface) method of 2.21. The water use efficiency was 180.85 kg/ha-mm with drip fertigation method compared to 59.53 kg/ha-mm for surface method.

TABLE 3.16   Benefit–Cost Analysis of Sugarcane (Maharashtra State, India)

	Cost economics	Micro irrigation system	Conventional system
1.	Fixed cost, Rs. per ha	30,000 (65,000 to 70,000 at present)	NIL
	a) Life, years	5	NIL
	b) Depreciation, Rs.	6,000	NIL
	c) Interest, @ 12%	1,800	NIL
	d) Repairs and Maintenance	600	NIL
	e) Total = (b)+(c)+(d); Rs.	8,400	NIL
2.	Cost of cultivation, Rs./ha	11,445	17,375
3.	Seasonal total cost = (1e) + (2); Rs/ha	19,845	17,375
4.	Water used, mm	940	2,150
5.	Yield of produce, 100 Kg/ha or tons/acre	1,700 or 68	1,280 or 51
6.	Selling Price, Rs. /100 Kg	31	30

	Cost economics	Micro irrigation system	Conventional system
7.	Income from Produce = (5) × (6); Rs.	52,700	38,400
8.	Net seasonal income= (7)–(3); Rs.	32,855	21,025
9.	Additional area cultivated due to saving of water, ha	2	NIL
10.	Additional Expenditure due to additional area = (3) × (9)	39,690	NIL
11.	Additional Income due to additional area = (7) × (9)	105,400	NIL
12.	Additional Net income = (11)–(10); (Rs.)	65,710	NIL
13.	Gross cost of Production = (3) + (10); Rs	59,535	17,375
14.	Gross income = (7) + (11); Rs.	158,100	38,400
15.	Gross benefit cost ratio = (14) ÷ (13)	2.66	2.21
16.	Net extra income due to micro irrigation system over conventional = (12) + (8, drip) – (8, conventional)	77,540	NIL
17.	Net profit per mm of water used = (8) ÷ (4)	34.95	9.78
18.	Water use efficiency [(5) ÷ (4)] × 100; Kg/ha-mm	180.85	59.53

Source: Sivanappan, R.K., 1994. Micro irrigation in India. Indian National Committee on Irrigation and Drainage (INCID), New Delhi, July.

Similar studies were carried out in Tamil Nadu, comparing the cost of cultivation for sugarcane under surface irrigation, subsurface drip fertigation (SSDF) and SSI under SSDF system (Table 3.17).

TABLE 3.17 Benefit–Cost Analysis of Sugarcane (Tamil Nadu)

S. No.	Particulars	Surface irrigation	SSDF	SSIDF
		Indian Rupees, Rs.		
1	Micro irrigation system cost (discounted cost), Rs. per year per ha	0	12,000	12,000
2	Preparatory cultivation	8,800	10,950	10,950
3	Setts and plating	23,640	16,160	15,500
4	Crop maintenance	8,400	8,400	8,400
5	Fertilizers cost	5,250	5,250	5,250

S. No.	Particulars	Surface irrigation	SSDF	SSIDF
		Indian Rupees, Rs.		
6	Irrigation/Drip fertigation	4,200	3,000	3,000
7	Weeding	3,000	3,000	3,000
8	Plant protection	2,470	5,060	5,060
9	Herbigation	0	1,257	1,257
10	Chlorine treatment	0	600	600
11	Acid treatment	0	800	800
12	Micronutrients	900	900	900
13	Harvesting	42,000	48,750	48,750
		Manual	Mechanical	Mechanical
14	Cane yield, tons/ha	98	175	195
	Economics			
15	Gross income, @ Rs.1,950/ton	19,1100	34,1250	390,000
16	Cost of cultivation, Rs.	98,660	116,127	115,467
17	Net income, Rs.	92,440	225,123	264,783
18	Benefit—cost ratio, BCR	1.93	2.93	3.29

SSDF = Subsurface drip fertigation, and
SSIDF = SSI under SSDF system.

The BCR for drip system was worked out by interviewing the farmers in Maharashtra and Tamil Nadu states by Sivanappan and his colleagues at TNAU [20–29]. The range of BCR excluding the proposition of water saving was from 1.31 to 2.60 for various crops excluding grapes, and for grape it was about 13.35. If water saving is considered, the BCR range goes up from 2.78 to 11.05 for various crops and 30.00 for grapes, respectively (Tables 3.18 to 3.20). This accounts for economic logic of entrepreneurial grape farmers to go in for the drip system in an expanded scale throughout India. Table 3.21 shows water productivity gains from shifting to drip irrigation from surface irrigation in India.

The research studies, at various research institutions in India, have indicated that the water saving for any crop is about 40–70% and the crop yield was increased up to 100% (i.e., double the yield). In spite of high installation cost of MI system, the economics were worked out by Sivanappan and his colleagues for various crops. They found that MI is viable. The payback period varies from 6–24 months and the BCR ratio is about 2.0 to 7.0.

TABLE 3.18 Benefit–Cost Ratio (BCR) for Various Crops Under Drip Irrigation in India

Crop	Row spacing m × m (ft. × ft.)	Cost of the MI system Rs/acre	BCR Excluding water saving	BCR Including water saving
Coconut	7.62 × 7.62(25' × 25')	7,000	1.41	5.14
Grapes	3.04 × 3.04(10' × 10')	12,000	13.35	32.32
Grapes	2.44 × 2.44(8' × 8')	16,000	11.50	27.08
Banana	1.52 × 1.52(5' × 5')	18,000	1.52	3.02
Orange	4.57 × 4.57(15' × 15')	9,000	2.60	11.05
Acid lime	4.57 × 4.57(15' × 15')	9,000	1.76	6.01
Pomegranate	3.04 × 3.04(10' × 10')	12,000	1.31	4.04
Mango				
Papaya	7.62 × 7.62(25' × 25')	7,000	1.35	8.02
Sugarcane	2.13 × 2.13(6' × 6')	18,000	1.54	4.01
Vegetables	Between lateral (6')	20,000	1.31	2.78
	Between lateral (6')	20,000	1.35	3.09

TABLE 3.19 Benefit-cost ratio (BCR) and Payback Period For Various Crops Under Micro Irrigation, India

Crops	Crop spacing m	Cost of the system Rs./ha	Water used Lpd/plant	Yield Tons/ha	Payback period months	BCR
Banana	0.91×1.5×1.8 pair row	47,500	15–20	75	12	3.00
Grape	3.03×1.8	44,000	15–20	45	<12	3.28
Pomegranate	4.3×4.3	30,000	50–60	25	<12	5.16
Ber	4.5×4.5	30,000	60	25	12	4.56
Tomato	0.45×.45×1.65 paired row	30,000 Cane-wall	40,000 (Lpd/ha)	75	One season, 6 Months	1.09
Papaya	1.81×1.81	40,000	15	60	12	4.09
Cotton	0.9×1.5×1.8 pair row	47,500	8–10	1.5	18	1.83
Sugarcane	0.83×1.66 pair row	47.500	30,000 Lpd/ha	200	12	3.45

SOURCE: Case studies conducted by the author R.K. Sivanappan with numerous farmers in Maharashtra State, November 1993.

TABLE 3.20   Benefit–Cost Ratio For Various Fruit Crops Under Micro Irrigation, India

Crops	Spacing m × m	Benefit-cost ratio, BCR	
		Excluding water saving	Including water saving
Grapes	3 × 3	13.35	32.32
	8 × 8	11.50	27.08
Acid lime	4.57 × 4.57	1.76	6.01
Banana	1.52 × 1.52	1.52	3.02
Mango	7.62 × 7.62	1.35	8.02
Orange	4.57 × 4.57	2.60	11.05
Papaya	1.84 × 1.84	1.54	4.01
Pomegranate	3.04 × 3.04	1.31	4.04

SOURCE: Constraints and potential in popularizing Drip Irrigation, R. K. Sivanappan and Associates, 1990.

TABLE 3.21   Water Productivity Gains From Shifting to Drip Irrigation From Surface Irrigation, India

Crop	Change in yield	Change in water use, %	Change in water productivity
Banana	+52	-45	+173
Cabbage	+2	-60	+150
Cotton	+27	-53	+169
Cotton	+25	-60	+255
Grapes	+23	-48	+134
Potato	+46	-0	+46
Sugar cane	+6	-60	+163
Sugar cane	+20	-30	+70
Sugar cane	+29	-47	+91
Sugar cane	+33	-65	+205
Sweet Potato	+39	-60	+243
Tomato	+5	-27	+49
Tomato	+50	-39	+14

SOURCE: 1. Sandra Postel (1999). Pillar of sand.
2. R. K. Sivanappan (1994). Data by Indian National Committee on Irrigation and Drainage: Drip irrigation in India. INCID, New Delhi.
3. R. K. Sivanappan (1994). Prospects of micro irrigation in India. Journal of Irrigation and Drainage Systems, 8:49–58.

Dr. Suresh Kumar at TNAU (2014) wrote a detailed report titled *Adoption of Sustainable Micro Irrigation in India: Factors and Policies* (Chapter 1, In: Management, Performance, and Applications of Micro Irrigation Systems, Volume 4, edited by M R Goyal. Apple Academic Press, Inc., 2015). Tables 3.22 and 3.23 summarize his findings on economics of crop production for banana and coconut. The economics of coconut cultivation in drip and control village revealed that the cost saving due to reduction in labor is 63%. Similarly, the cost of cultivation has considerably reduced under drip method registering a reduction of 9.1%. It is interesting to note that the drip method resulted in high water and energy productivity.

TABLE 3.22 Economics of Crop Production (Rs. per ha) for Banana in Sample Farms, 2007–2008

Particulars	Drip adopters	Non-adopters
Quantity of water pumped (M3)	8506.3	21,316.9
Quantity of energy consumed (kwh)	2670.9	7313.9
Cost of labor (Rs.)	11,123.4***	25,075.4
Capital (Rs.)	70,678.3***	94,752.2
Yield (quintals)	605.6	591.5
Gross income (Rs.)	259,937.5	254,230.8
Gross margin (Rs.)	189,259.2***	159,478.5
Yield per unit of water (Kg/M3)	7.1***	2.8
Yield per unit of energy (Kg/kwh)	22.5***	8.3
Returns per unit of water (Rs/M3)	21.8***	7.6
Returns per unit of energy (Rs/kwh)	68.1***	22.9

***, ** and * indicate values are significantly different at 1%, 5% and 10% levels from the corresponding values of control village.

The analysis of economics of crop cultivation under drip and flood methods revealed that the drip method of irrigation has significant impact on resources saving, cost of cultivation, yield of crops and farm profitability (Tables 3.14–3.23). The physical water and energy productivity is significantly high in drip method of irrigation over the flood method of irritation.

TABLE 3.23 Economics of Crop Production (Rs. per ha) for Coconut in Sample Farms, 2007–2008

Particulars	Drip adopters	Non-adopters
Quantity of water pumped (M3)	13,185.5	21,584.7
Quantity of energy consumed (kwh)	905.2	5774.9
Cost of labor (Rs.)	4670.1***	12,463.5
Capital (Rs.)	29,814.4***	32,798.3
Yield ('00 nuts)	231.8***	199.4
Gross income (Rs.)	113737.3	85,084.2
Gross margin (Rs.)	83,922.8	66,145.8
Yield per unit of water (nuts/M3)	1.8***	1.0
Yield per unit of energy (nuts/kwh)	25.9***	3.8
Returns per unit of water (Rs/M3)	6.5***	3.4
Returns per unit of energy (Rs/kwh)	95.5***	12.6

***, ** and * indicate values are significantly different at 1%, 5% and 10% levels from the corresponding values of control village.

3.8 DESIGN OF DRIP IRRIGATION SYSTEM

For in depth study, reader should consult following two chapters:

1. Wu, I. Pai and H. M. Gitlin, 2013. Design of trickle irrigation systems. Chapter 12, pages 219–246, In: *Management of Drip/Trickle/ or Micro Irrigation,* edited by M. R. Goyal. New Jersey, USA: Apple Academic Press, Inc.

2. Wu, I. Pai and A. L. Phillips, 2013. Design of lateral lines. Chapter 13, pages 247–259, In: *Management of Drip/Trickle/ or Micro Irrigation,* edited by M. R. Goyal. New Jersey, USA: Apple Academic Press, Inc.

These two chapters by Wu [30, 31] describe in detail the methods for design of micro irrigation systems. Authors of these two chapters have included solved examples, design charts, and nomographs. Dr. Megh R. Goyal, Senior Editor-in-Chief of Apple Academic Press Inc. has also published following book consisting of 16 chapters:

Goyal, M. R. and M. A. El-Nesr, 2014. Sustainable Micro Irrigation Design Systems for Agricultural Crops: Practices and Theory. New Jersey, USA: Apple Academic Press, Inc.

The design of a drip irrigation system is based on the hydraulics of pipe flow. For designing the drip layout for any crop, the irrigation designer must have the following information:

- The water source: its elevation head or pressure at which water is available.
- Types of crop and agronomical practices.
- Topography of land and size of field.
- The soil characteristics: its permeability and infiltration rate.
- Climatic data.

The drip irrigation system is based on watering plants individually at frequent intervals. The design includes: design of emitters, laterals, submains, mains, filtration system, chemigation system, automation (optional), economic analysis, etc., respectively. The irrigation systems can be permanent or portable, and surface or subsurface.

To calculate the daily water consumption of the plant (CU), we need USDA Class A pan evaporation data during the hottest period at the site.

$$CU = Epan \times Kc \times S \times R \tag{6}$$

where: CU = Daily water consumption of the plant, mm/day; Epan = USDA Class A pan evaporation, mm; Kc = Crop coefficient depending on the growth period, a fraction; S = Plant spacing, m; and R = Row width, m.

3.8.1 EMITTER SELECTION

The number of emitters required at each location of a plant is based on the soil-wetted volume in the vicinity of a plant. The nature and development of roots must be known to select the minimum volume of wetting of soil. One must ensure proper anchorage for the plant. The soil structure and texture affects the wetted area and volume. Research studies have indicated that the emitters must be able to wet 40 to 60% of the wetted or shaded area to avoid water stress. If a single emitter is provided for each plant, it must be placed 15 to 30 cm from the base of the plant.

$$N = [A/(\pi r^2] \tag{7}$$

$$E = [q/nt] \tag{8}$$

where: N = the number of emitters by Eq. (7); A = total area to be wetted; and r = radius of wetted area for a single emitter; E = flow rate of each emitter by Eq. (8); q = daily water required; n = number of emitters; and t = hours of operation per day.

3.8.2 IRRIGATION INTERVAL

The irrigation interval depends on the quantity of water in each irrigation and the daily requirement of a plant. Karmeli and Keller [7] calculated the irrigation interval as follows:

$$I_i = [I_{dn}/T] \tag{9}$$

$$I_t = [(KI_d S_e S_l)/q_a] \tag{10}$$

where: I_i = Irrigation interval Eq. (9), hours; I_{dn} = Net depth of each irrigation, mm; I_t = time of operation of each emitter Eq. (10), hours; K = conversion constant = 1 (metric system); S_e = Emitter spacing, m; S_l = Average lateral spacing, m; and q_a = Average emitter discharge, lph. The time of operation of each emitter is selected to avoid any excess runoff or percolation. The emitter discharge rate and soil infiltration rate influence the time of each irrigation.

3.8.3 EMITTER DESIGN

In the drip irrigation system, the emitters must be designed to provide water at operating design pressure of the system. The performance of the emitters can vary with the flow type through the emitter. If the flow is laminar, it can cause clogging. If the flow is turbulent, the opportunity of clogging is less. The uniformity of drip irrigation system is dependent on the flow characteristics of the emitter, emitter manufacturing tolerance, emitter uniformity coefficient, manufacturer's coefficient of variation and the permissible pressure variation in the system. The emitter must satisfy the following requirements for achieving acceptable uniformity [7]:

- They must provide low but uniform and constant discharge, which does not vary significantly due to minor difference in pressure.
- They may have relatively large flow cross section to reduce clogging problem.
- Be inexpensive, compact and accurately made.
- Be locally available.

We want less discharge at a high-pressure drop, under Indian conditions. However, enlarging the area of emitter cross section will result in less pressure drop. These two requirements are contradictory. This had led to the development of a number of drippers throughout the world. Today, there are more than 400 types of emitters that can be used in different situations and for various crops. Karmeli and Keller [7] indicate following design equation for an emitter flow:

$$q = K_d H^x \tag{11}$$

Taking log of both sides, we get:

$$Log\ q = Log\ K_d + \times Log\ H \tag{12}$$

where: q = emitter discharge, lph; K_d= constant for a specific emitter; H = Operating pressure head; and $\times$ = Exponent which is characterized by a flow regime. From a plot of (Log q) versus (Log H), we can values of K_d and x. The reader will note that Eq. (12) is a straight line. One can therefore plot on an ordinary graph paper. Value

× is a slope of line defined by Eq. (11) and Log K_d is an intercept. Taking antilog of this intercept will give a value of K_d. For a fully turbulent flow, $x = 0.5$; for a partially turbulent flow, $x = 0.5$ to 0.7; for the unstable flow regime, $x = 0.7$ to 1.0; for laminar flow, $x = 10$; and for microtubes, $x = 1.0$

The emitter discharge is a function of temperature and pressure. Where calibrations were made with a water temperature of 20 °C, the discharge is multiplied by the following factors (assuming the same pressure head and laminar flow):

θ, °C	5	10	15	20	25	30	35	40
Factor	−0.63	−0.75	−0.87	−1.00	−1.13	−1.28	−1.43	−1.56

This head loss (friction drop) between two emitters can be calculated by:

$$h_i = f[L/D]\,/[v^2/2\ g] \tag{13}$$

where: h_i = head loss or losses due to friction, m; f = coefficient of friction depending on Reynolds' number; L = length of pipe; v = mean velocity; g = acceleration due to gravity = 32.2 fps = 9.81 m/s²at the surface of earth; d = pipe diameter. The orifice discharge is defined as:

$$q = AC_d[2\ gh]^{0.5} \tag{14}$$

where: q = orifice discharge; C_d = coefficient of discharge; A = orifice area = $[\pi(r^2)]$; r = orifice radius; g = acceleration due to gravity = 32.2 fps = 9.81 m/s² at the surface of earth; and h = pressure head at the orifice. If we consider the flow through a micro –tube as an orifice flow, then the head loss for an emitter (a micro – tube type emitter) can be found by Darcy-Weisbach equation:

$$h_f = f\,[L/D][v^2/2\ g] = f\,[8Lq^2]/[gD^5\pi^2] \tag{15}$$

where: h_f= head loss; D = diameter of a microtube; f = coefficient of friction depending on type of flow; L = length of microtube; and q = discharge of emitter; and g = acceleration due to gravity = 32.2 fps = 9.81 m/s² at the surface of earth; and 8 = constant. The Darcy–Weisbach equation is exact for laminar flow and can be derived theoretically. The formula may be extended to turbulent flow by varying the coefficient of friction, f. The Colebrook-White equation for the turbulent friction factor was derived based on experiment. Graphs for emitter flow in lph versus operating pressure head in meters are available from the manufacturer for a particular type of emitter. One can also develop such a graph from the experimental data. At a selected pressure, a known amount of water sample is collected for a fixed time. Data is converted into lph. By using Eqs. (11) and (12), we know the constants for a specific type of emitter. Sivanappan and his colleagues tested different types of emitters at TNAU and developed the flow characteristic curves for two types of nozzles, hole and socket type of emitter (1 mm in diameter), and microtube (1 mm in diameter). They found the relationships to be nonlinear. Although an empirical equation was not developed, yet the shape of each curve is similar reported by other investigators [7, 31, 32]. A typical curve is shown in Fig. 3.24.

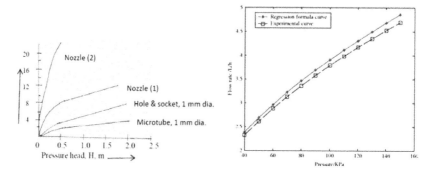

FIGURE 3.24 A typical flow rate in lph (Y-axis) versus pressure (X-axis) curve for an emitter.

3.8.4   DESIGN OF LATERALS

In drip irrigation system under Indian conditions, the laterals are low cost PVC, PE or alkathene pipes. The flow condition in the lateral is steady and is spatially varied with lateral outflow. They are designed to carry uniform discharge and supply water through the emitters with acceptable uniformity. The lateral size should be selected to carry the maximum water required for one row per unit time. The slope of the lateral line affects the discharge through the emitters, as it causes pressure change in the line. It should also be taken into account that water is discharged through the various emitters all along the lateral. In the drip laterals, the pressure drop between the lateral lines must not exceed 20 percent of the emitter operating pressure. When there are fixed number of emitters, the length of laterals is determined by the pressure drop between the lateral lines or the uniformity of emitters discharge. In the drip irrigation system, the emitter uniformity must not be less than 90 percent [7].

$$E_d = [100 - (1 - \Delta q)/q_{avg}] \geq 90\% \qquad (16)$$

where: Ed = distribution efficiency, %; Δq = discharge variation; and q = average discharge; The pressure distribution along the laterals is determined by the pressure drop due to friction, pressure gain for down slope, and pressure drop for up slope situations. The profile of emitter flow along a lateral line will be similar to the pressure distribution in the lateral line. There are 3 profiles that are discussed by Wu [7, 30, 31]:

1. The emitters flow decreases with respect to the lateral length. This occurs when the lateral line is laid on level ground or up slope. In this condition, the Q move is determined by the operating pressure.

2. The emitter flow decreases with respect to the lateral length and reaches a minimum emitter flow point and then increases with further length of lateral line. This occurs when a gain of energy by slopes at downstream point is larger than the energy drop due to friction. This type occurs when the lateral line is laid on mild downhill slopes.

3. The emitters flow increases with respect to the lateral length. This is caused by steep slopes, when the energy gain is larger than the friction losses for all sections along the lateral line. In this condition the minimum value is determined by input pressure.

For these profiles, the relationship is nonlinear between application efficiency and emitter flow variations. The increase in emitter flow variation causes decrease in the application efficiency are plotted as shown in Fig. 3.25. The distribution efficiency is relatively high for drip irrigation since the water distribution is under full control. The distribution efficiency varies inversely with the emitter flow variation and the relationship is curvilinear.

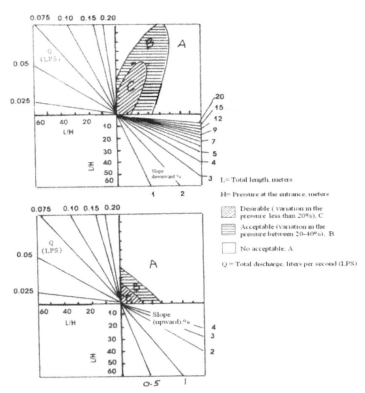

FIGURE 3.25   Design chart for 12 mm lateral line. Top figure: slope downward, and bottom figure: slope upward.

3.8.4.1 DESIGN PROCEDURE FOR LATERAL LINE

Step 1: Determine water use, coverage factor, length of laterals, time of operation and maximum allowable pressure variation in the laterals.

Step 2: Determine the design application rate as follows:

$$DR = CU \times A \times K \tag{17}$$

where: CU = peak consumption per day; A = plant space area; and K = Coverage factor of a crop.

Step 3: Determine the number of emitters, emitter flow rate, lateral pressure, emitter spacing and the emitter type.

Step 4: Determine the operating time.

Step 5: Choose an initial lateral pipe size.

Step 6: Determine the lateral pressure drop and uniformity coefficient. If they are not within the allowable range, change the pipe size and repeat the procedure.

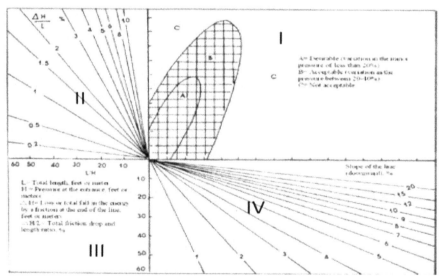

FIGURE 3.26 Dimensionless design chart for lateral and secondary lines (downward of the slope).

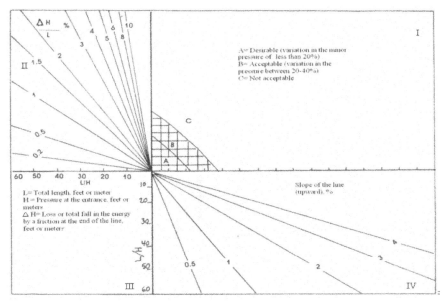

FIGURE 3.27   Dimensionless design chart for the lateral and secondary lines (upward of the slope).

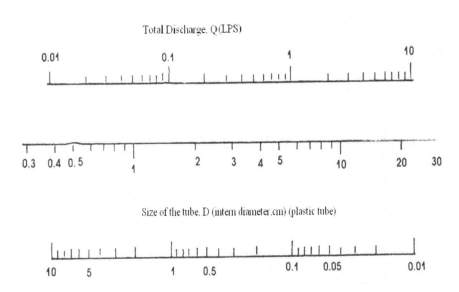

FIGURE 3.28   Nomograph for the design of lateral and secondary lines in metric units.

Alternative method has been suggested by Wu and his colleagues [7, 30, 31] of University of Hawaii. They have developed design charts based on the basic hydraulics of drip irrigation lines and computer simulation. The charts can be used for any specific diameter of the lateral line. The Figures 25 to 28 are for general design [7]. The procedure is described below [7].

Step 1: Establish along one of the lateral lines: lateral length L and operating pressure H, ratio (L/H) and the total discharge in liters per second (lps).

Step 2: Move vertically from L/H in quadrant III to the given total discharge (lps) line in quadrant II. Then establish a horizontal line toward quadrant I.

Step 3: Move horizontally from L/H in quadrant III to the % slope line in quadrant IV. Then establish a vertical line toward quadrant I.

Step 4: The point of intersection of these two lines in quadrant I determines the acceptability of the design.

If desirable pressure variation is less than 20%, then emitter flow variation is less than 10%. Any pressure variation larger than 40% and emitter flow variation larger than 20% is not recommended.

Another method known as poly plot method was developed by Elaine Herbert and Lesculumn of ICI computer group in Australia. The energy loss in a lateral line can be calculated by using the William and Hazen formula [7, 31, 32].

$$H = 5.35(L)[Q^{1.852}/D^{4.871}] \qquad\qquad (18)$$

where: H = energy loss by friction (m) at the end of lateral or submain; Q = total discharge (lps) of lateral or submain; D = inside diameter of the lateral (cm); and L= length of pipe section of lateral (submain) in meter. As the discharge in the lateral or submain decreases with respect to the length, the energy gradient line will not be a straight line but a curve of exponential type. However, it can be expressed by a dimensionless energy gradient line. When the lateral is laid on slopes, the pressure variation can be determined as a linear combination of energy slope and line slope (the change of velocity head in the line being small is neglected). This is expressed as:

$$[dh/dt] = - S_f \pm S_o \qquad (7) \qquad\qquad (19)$$

where: [dh/dt] = dimensionless energy gradient line; S_f = slope of energy gradient line; S_o = line slope; and ± for up or down slope. An allowable pressure variation can be set along the energy gradient line and can be drawn as a curve. The area between the energy gradient line and the allowable pressure variation curve can be used for designing drip irrigation lines for both uniform and nonuniform slopes. If the line slope can be put within the area, the design will have a pressure less than the set allowable pressure variation. This concept is proposed as "Poly Plot."

The procedure of using the poly plot method is listed as follows:

Step 1: Select a lateral line

Step 2: Determine Q100 for the operating pressure. Q = Total discharge in lps of lateral/submain.

Step 3: Trace an energy gradient line for the determined Q100 on the work sheet.

Step 4: Set up allowable variation of pressure. Plot the allowable variation of pressure curve on the worksheet.

Step 5: Plot the lateral line slope profile on a transparent paper. The plot should have the same scale as the energy gradient line.

Step 6: Put the lateral line slope profile on the worksheet as determined in **steps 3 and 4** and match the end point of the profile to the zero lateral length.

Step 7: Move the transparent paper up and down and superimpose the lateral line slope profile on the area between the energy gradient line and the allowable variation of pressure curve.

Step 8: If the lateral line slope profile can be put inside the area, the design has a pressure variation less than the allowable variation of pressure.

Step 9: If the lateral line slope profile cannot be put inside the area, the design has a pressure variation more than the allowable variation of pressure.

3.8.5 DESIGN OF SUBMAINS

The design of submain is similar to design of a lateral line. The total flow in the lateral lines is considered as outflow from the submain. The average discharge is good estimate to calculate the energy loss along the line. However, adding 20% to the calculated value will be close to the actual friction loss in the submain.

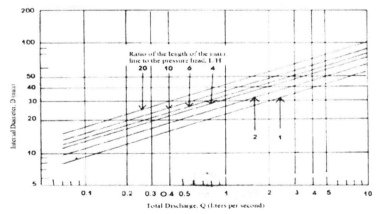

FIGURE 3.29 Design chart for the secondary line for slope less than 0.5% and allowable variation in pressure of 10% (M.K.S. units).

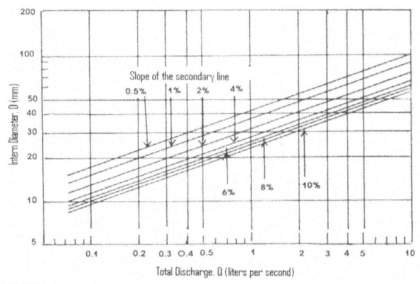

FIGURE 3.30 Design chart for the secondary line for slopes equal or greater than 0.5% (M.K.S. units).

3.8.5.1   DESIGN PROCEDURE:

The submain length is determined by the number of laterals served and the distance between laterals. Separate charts are prepared for lines, the slope of which is less than 0.5% and slope greater than 1.5%. The results are shown in Figs. 3.29 and 3.30.

Step 1: Determine the total discharge Q for submain.

Step 2: Determine the length and pressure head ratio L/H.

Step 3: Determine the submain slope. If the slope is equal or greater than 0.5%, use Fig. 3.30 to design submain size.

Step 4: If slope is less than 0.5% then curves are to be drawn operating for it and then use it to get the pipe diameter (Fig. 3.29).

Step 5: Since we know L/H ratio, we can make use of the Nomograph for total friction drop H and length (L), ratio L/H, total discharge Q (lps) to find the required size of pipe.

3.8.6   DESIGN OF MAIN LINES

The design of a drip irrigation system is based on the hydraulics of pipe flow. The system should be designed to meet the crop irrigation requirements to have enough capacity: To provide extra irrigation during unusually dry periods; to distribute water into fields with acceptable uniformity; and to have the lowest cost among other design alternatives.

The main line design is based on the topography of the field, the operating pressure, the field layout of laterals and submains, and the required discharge from each outlet along the main line. The main line system therefore has a variable flow capacity (discharge in the pipe) with respect to the length. It has more discharge at the up-steam sections than in the down-stream sections. The design of main involves selection of a proper pipe size for each section to deliver water at the required rate to all the submains in the system. There are numerous arrangements of pipe sizes that can be designed to meet the hydraulic situation of a given layout. There are many different layouts and each means a different cost (See Section 3.7). The final objective of designing a main line is not only the optimal design within a given field layout, but also the optimal design among several field layouts.

3.8.6.1   ENERGY GRADIENT LINE METHOD

The main line design is a series of pipe flow designs. Once the field layout is set, the discharge rate in each section can be determined. The Williams and Hazens formula is used to determine pipe size [7, 31, 32]:

$$H = \{15.27L[Q^{1.852}]\}/[D^{4.871}] \tag{20}$$

where: H = the energy loss head for a given length L, in meters; L = length of a pipe; Q = discharge rate in lps; and D = the inside diameter of the pipe in cm. Equation (20) can be rearranged in terms of energy slope, H/L, dimensionless.

$$[H/L] = \{15.27[Q^{1.852}]\}/[D^{4.871}] \tag{21}$$

Using Eq. (21), the pipe size D can be calculated for a given discharge Q, if the energy slope is known. This means that there are numerous solutions of pipe size D for different specified energy slopes for a given discharge Q. The energy slope should be selected in such a way that the energy gradient line is above the required water pressure along the main line. Therefore, the water pressure in the main line will be always equal to or higher than the required water pressure for flow along the main line [7, 31, 32].

3.8.6.1   DESIGN PROCEDURE FOR MAIN LINE

Step 1: Plot the main line profiles and required pressure for drip irrigation operation as shown in Fig. 3.31.

Step 1: Draw a straight energy gradient line from the available operating pressure (Y-axis) to the required pressure profile, so that everywhere along the main line, the energy gradient line is above the required pressure profile.

Step 1: Determine the energy slope, which is the slope of straight energy gradient, ΔH/L.

Step 1: Determine required discharge for each main line section.

Step 1: Design main line size by using nomograph in Fig. 32.

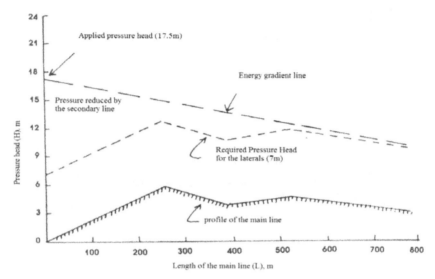

FIGURE 3.31   Profile of energy gradient line and profile of the main line (M.K.S. units).

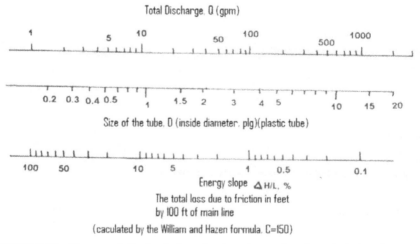

FIGURE 3.32   Nomograph for the design of main and secondary lines (M.K.S. units).

For maximum flow rates at minimum friction loss and laminar flow, size of PE pipes is shown in Table 3.24.

TABLE 3.24   Pipe Size For Different Flow Rates

Diameter of pipe, D	Flow Rate, Q
mm	lpm
19.05	6.91
25.40	11.40
31.75	17.00
38.10	25.00
50.80	45.50
76.20	100.00

3.8.7 HEAD WORKS

3.8.7.1. PUMP AND MOTOR

Head works or pump station includes pump with a motor, filtration system, fertigation tanks with injectors, pressure regulator and gate valves, etc. Once the diameter and discharge of all laterals, submain and main lines is determined, flow and pressure required for the pump are determined. If the motor is of high pressure type, it is essential to reduce the supply pressure. Two types of pressure reducing devices are usually used: Flow restricting device which reduce the water flow in pipe lines but which are not adjustable for different input pressures; and self-compensating device which supply water at a constant pressure. Fertilizer tanks are fitted in most of the drip irrigation system.

3.8.8 FERTIGATION

All soluble fertilizers can be fertigated by using this fertilizer tank [7]. Since the water is uniformly distributed, it is possible to apply fertilizer uniformly to each individual plant. It is also possible to achieve a 50% reduction in the quantity of fertilizer requirement. The fertigation procedure is described in detail Chapter 8 and 9 by Goyal [7]. After filling the tank to capacity with previously computed quantity of liquid or dissolved fertilizer, the tank is closed. Irrigation is then started with the regulating valve completely open and the inlet to the tank is still closed. Only after filling the system with water and allowing operation for a certain time, the inlet valve to the fertilizer tank is opened and fertilizing is started. Golden rule of thumb is to fertigate in the middle of irrigation duration. Simultaneously the regulating valve is set to the proper position.

3.8.9 FILTRATION SYSTEM

To avoid clogging, a good filtration system is a must [7, Chapters 9–11]. As the emitter openings are 1 mm, therefore irrigation water must be very clean free of any impurities and sediments. Partial clogging may result from sedimentation, precipitation of salts and residues, etc. The causes of clogging are: Solid particles in suspension, microorganisms, and chemical precipitation. To prevent the clogging, a filtration system must be adequately designed. The various types of filters are: Screen type strainers; gravel and filters; and hydro-cyclone filters (Table 3.25). Screen type strainers are of simplest type. It consists of two cylindrical screens. The outer screen is of 80 mesh screen and serves to remove the large particles. Inner screen is of 120 mesh screen. The units can be easily cleaned using a flushing system. Gravity filtration is not commonly used. The separation of solids from liquids by centrifuge is a commonly used method.

The gravel filter is relatively large in cross section and height and the flow of water is from top to bottom. The latest models have proved that gravel of only one size range (basalt) is sufficient. Cleaning is done by back flushing that involves allowing the flow from bottom to top of a filter, while the system is in operation.

The vortex filter is an inverted cone shaped hollow tank with a single inlet and an upward outlet. The sand sinks alongside the container walls due to the centrifugal force exerted during the vertical water flow. The sand at the apex bottom of the container is flushed from time to time, by opening the gate valve at the bottom of a tank [7].

Though expensive, compressed air or water is applied to clean the clogged emitters at a pressure of 700 cbars. Carbonate precipitates cannot be removed from the drippers by this method. One will need to use the chlorination method to remove Ca and Mg salts. Reader is advised to consult book by Goyal [7].

To reduce the clogging in the laterals and emitters, flushing is usually done frequently by allowing water flow freely at the open ends of laterals for a reasonable time. Chemicals such as hypochlorites are also found to be effective in improving the flow rates of partially clogged emitters [7]. The system will be evaluated for uniformity before being operated for the first time [10].

TABLE 3.25   Types of Filters in Drip Irrigation [7]

TYPE OF FILTER	APPLICATIONS	FUNCTION	SPECIFICATIONS	REMARKS
Sand Media	a. Required for any open or surface water sources where large amounts of organic matter are present. b. Frequently used with well water	Fine sand particles within two or more closed tanks create a three-dimensional filtering surface trapping algae, slimes and fine suspended solid particles. Tanks are back-flushed one at a time, while remaining units continue filtration.	a. Filtration from 200 mesh (74 µm) to 600 mesh (25 µm) b. Tank sizes are typically ranging from 12-48 in.-dia. (0.30-1.20 m). c Recommended using at least three tanks.	a. Cleaned by back-flushing. b. Stainless steel epoxy-coated or fiber-glass tanks are available for acid injection. c. Several tanks can be used in parallel for large flow rates.
Screen	a. May be used as a primary filter for clean water sources. b. Can be a safety back-up downstream from the sand media filter. c. Can be used as a submain secondary field filter	a. Fine mesh screen(s) enclosed in one or more pressurize tanks traps organic and inorganic particles. b. Filter can be cleaned manually or automatically by various high pressure rotating water jets and/or brushes	Available screen materials and mesh varies based on manufacturers and types of filter; common sizes: 50-200 mesh (300-74 µm).	a. Cleaned by manual removal or automatic flushing while using rotating water jets. b. can be easily clogged by organic contaminants
Disk	a. Use for primary filtration similar in application to media filters.	a. Filters through densely packed thin color-coded polypropylene disks that are grooved diagonally on both sides to a specific micron size (Fig. 7-28). b. The flushing process starts automatically when given pressure differentials or time setting are reached; flush commands from the controller.	a. Commonly available disks range in sizes: 18-600 mesh (800-25 µm). b. Multiple filter configuration adjustable to water quality and capacity demands.	a. Flushing commands are sent from the electronic controller. b. Flushing is rapid and water efficient. c. Stacked filter do not require a lot of space. d. Disks should be replaced annually unless not processing a lot of dirty water.
Gravity-Flow	a. Used for low levels of particulate matter. b. Used to deliver a large volume of water at low pressure.	Water falls on a screen separator which traps particulate matter which is then washed out into a collection tank	Available from 100 to 200 mesh (150-74 µm).	a. Cleaned by water flow and additional spray nozzles. b. Booster pump is usually necessary after this filter.
Centrifugal Sand Separator	a. Used to remove sand and other inorganic particles. b. Used as a pre-filter to help reduce back-flushing of main filter.	Centrifugal action creates a vortex that pushes away particles heavier tan water. Removes well casing scale, sand and other inorganic particles.	Removes particles heavier than water down to 200 mesh (74 µm). b. Works with a 5-7 psi (0.35-0.49 kg/cm²) pressure loss.	a. Self-cleaning. b. Low maintenance. c. Does not remove organic matter. d. Is not 100 % effective-usually used as a pre-filter.
Suction screen	For pre-filtration at pump intake in ponds or reservoirs or lakes.	Coarse screen traps debris, birds and fish. preserve foot-valve pump.	Available in 10-30 mesh (1500-500 µm).	Cleaned by rotating inner water jets
Settling basin	Pre-filter to remove silt or other inorganic particles.	Allows silt and clay particles to settle. May also provide aeration to remove dissolved solids and iron in suspension.	Sized based on peak water budget and particulates types and load	Cleaned by draining and removing build up. Inlet must be away from inlet. Must control algae.

3.9   SUMMARY

The ruthless and biased attitude in the selection of the farm irrigation system to be adopted in a project tends to retard and materially impede the irrigation progress. The research results, at different agricultural universities/ centers/ institutions in

India, show a direct relationship between crop response and moisture tension. It has also been attested that evapotranspiration is high in arid and semiarid regions of the world. Appendix I indicates water consumption and yield for various crops in drip and conventional irrigation methods in India.

With drip irrigation, the total initial cost per ha of the system is closely proportional to the number of distribution laterals, mains, submains, emitters, etc. for a specified area. There is tremendous scope for further research studies. The information in this chapter will surely help to provide an understanding of the basic principles involved in this novel method of water saving and increase in crop yield. Heavier soils make drip irrigation installation cheaper. Drip irrigation can supply small quantities of water at any desired interval of irrigation, near the plant. It offers a most unique moisture regime in the soil for better plant growth, which contributes to increase the yield.

The real problem is the ability to design a trickle network density based on optimal application requirements as well as on a sound benefit-cost ratio. Filtration of the water is most important challenge to avoid clogging of drippers and laterals. Agronomists fear that an irrigation system, which eliminates or reduces the phase of aeration of the soil between applications ratio has various disadvantages, which is nonexistent with conventional watering technique.

Higher acreage is possible for drip irrigation in India, if the initial cost of installation of drip irrigation is brought down so that small farmers can have easy access. Enough literature exists throughout the world on principles and design of drip irrigation system, in search for solutions to challenging tasks of reducing irrigation water for agricultural crops.

Due high degree of control in fertigation of fertilizers and automation, drip irrigation has proved of great benefit to farmers. Higher crop yields can be obtained by drip method. Reports from countries outside India indicate clearly that yield, quality and water use efficiency will definitely ensure that drip irrigation is an option available. It is a most accurate and sophisticated method of providing water for growing of crops. It is well suited in places where irrigation water is costly. This advanced method of drip irrigation will replace the age old surface flooding in the years to come especially when the water becomes scarce and costly. Let us all save the planet from diminishing water resources.

KEYWORDS

- alkathene tubes
- apparent density
- application efficiency
- available soil moisture depletion, ASMD

- benefit–cost ratio, BCR
- biwall drip tubing
- bottle guard
- button dripper
- Central Arid Zone Research Institute
- centrifugal pump
- compensating emitters
- consumptive use, CU
- cotton
- Darcy–Weisbach equation
- distribution efficiency
- drip irrigation layout
- dripper
- economics
- emitter
- evaporation
- evapotranspiration, ET
- fertigation
- field capacity
- grapes, table
- grapes, wine
- gravimetric method
- gypsum blocks
- hole puncher
- Indian Council of Agricultural Research, ICAR
- irrigation requirement, IR
- laminar flow
- lateral
- main line
- musk melon
- net returns
- neutron scattering method
- New South Wales, NSW
- orchard crops
- pan evaporation, Epan
- pepper
- plantations
- poly vinyl chloride, PVC
- polyethylene, PE
- relative humidity
- ridge guard

- **seep hose**
- **self-flushing emitters**
- **soil moisture**
- **submain**
- **subsurface drip irrigation, SDI**
- **sugarcane**
- **surface drip irrigation**
- **sweet pepper**
- **Tamil Nadu Agricultural University, TNAU**
- **tomato**
- **transpiration**
- **turbulent flow**
- **twin wall drip tubing**
- **uniformity coefficient**
- **USDA Class a pan**
- **water use efficiency, WUE**
- **wilting point**

REFERENCES

1. Cole, P. J., Armstrong, D. W. Design and operation of drip irrigation systems for South Australian conditions. Report 2/73 by Dept. of Agriculture, Australia.
2. Cole, P. J. *Trickle Irrigation*. Progress Report on Demonstration Trials at Golden Heights, Dept. of Agriculture, Australia.
3. Department of Agriculture. Daily flow irrigation of South Victoria Province. Publication of the Dept. of Agriculture, Australia.
4. Goldberg, D. *Modern Concepts on Irrigation*. Hebrew Uni., University of Jerusalem.
5. Goldberg, D., Ben Asher and B. Gosret, (1976). Soil and water status under sprinkling and trickling. Agricultural Water Management, 1, 33–40.
6. Goldberg, D., M. and Shamueli, (1970). Drip irrigation: A method used under arid and desert conditions of high water and soil salinity Trans. American Society of Agri. Engineers, *13(1)*.
7. Goyal, Megh R., (2013). *Management of Drip/Trickle or Micro Irrigation*. New York: Apple Academic Press, Inc. Pages 426.
8. ICAR, 1979. Annual Progress Report: 1978 and 1979. Indian Council of Agricultural Research, New Delhi.
9. *Irrigation and Drainage Paper on Trickle Irrigation*. FAO publication, Rome, Italy. 1973.
10. *Methods of Evaluating Irrigation Systems*. Hand book 82 by U. S. Department of Agriculture (USDA), Washington, DC.
11. Padmakumari, O., Sivanappan, R. K. Economics of drip irrigation for water scarce regions. MAJ (India), *66(7)*, 486–490.
12. Padmakumari, O., Sivanappan, R. K. Drip irrigation in brinjal: Its economics and adoptability in water scarce areas. MAJ (India), *65(9)*.
13. Padmakumari, O., Sivanappan, R. K. (1980). *Drip Irrigation principles and Practices*. Kisan World, 7(3), 17–19.

14. Padmakumari, O., Sivanappan, R. K. (1977). An advanced irrigation method for better yield in cotton. MAJ, 66(3), 174–176.
15. Padmakumari, O., Sivanappan, R. K. Wetting pattern for varying rates of dripper discharge. MAJ, 66(4), 271–272.
16. Proceedings of the International Conference on drip irrigation, 1975. San Diego USA.
17. Shamuali, M., Goldberg, D., Gornat, B. (1967). The influence of the irrigation and soil treatments on sprouting of onion under the Arava Desert conditions. Ketavim Journal, 17(4).
18. Shamuali, M., Goldberg, D. (1969). Advances in trickle irrigation. Proceedings of Israel Soil Science Soc.
19. Singh, S. D., Niwas, R. P. S. (1973). Trickle irrigation ensures higher water use efficiency. Indian Farming, October.
20. Sivanappan, R. K. (1977). Sprinkler and drip irrigation. Indian Farming May.
21. Sivanappan, R. K. Drip irrigation. Irrigation Era, 14(4), 19–24.
22. Sivanappan, R. K., Natarajan, P. (1974). Studies on drip irrigation method in Tomato. Madras Agri. Journal (MAR), 61(9), 888–890.
23. Sivanappan, R. K. (1977). Drip irrigation for fruit crops. Paper presented at International symposium on citrus culture organized by Horticultural Society of India, Bangalore.
24. Sivanappan, R. K. (1977). Economics of drip irrigation method in small and medium farms. Paper published in the Souvenir of ISAE XV Annual Convention February, pages 27.
25. Sivanappan, R. K. (1976). Drip Irrigation in banana. Indian Farming, New Delhi, July.
26. Sivanappan, R. K. (1974). The response of bendhai to the drip system of irrigation. South Indian Horticulture, 22(3 & 4), 95–100.
27. Sivanappan, R. K. (1976). Method of efficient water use by drip irrigation. Journal ISAE, pages 37.
28. Sivanappan, R. K. (1977). Trickle irrigation for water scarcity areas. The Agricultural Society Engineer, 20:23–25.
29. Sivanappan, R. K., Gowder, K. R., Gandhi, M. (1972). Drip Irrigation. MAJ, 50, 440–441.
30. Wu, I. Pai; Gitlin, H. M. (2013). Design of trickle irrigation systems. Chapter 12, pages 219–246, In: *Management of Drip/Trickle/ or Micro Irrigation,* edited by M. R. Goyal. New Jersey, USA: Apple Academic Press, Inc.
31. Wu, I. Pai; Phillips, A. L. (2013). Design of lateral lines. Chapter 13, pages 247–259, In: *Management of Drip/Trickle/ or Micro Irrigation,* edited by M. R. Goyal. New Jersey, USA: Apple Academic Press, Inc.

APPENDIX I

Water Used and Yield For Various Crops in Drip and Conventional Irrigation Methods in India

Crop	Yield (100 Kg per ha)			Water depth applied (cm)		
	Conven-tional	Drip	Increase in yield (%)	Conven-tional	Drip	Water Saving (%)
Banana	575.00	875.00	52	176.00	97.00	45
Grapes	264.00	325.00	23	53.20	27.80	48
Mosambi	100.00	150.00	50	166.00	64.00	61
Pomegranate	55.00	109.00	98	144.00	78.50	45
Sugarcane	1280.00	1700.00	33	215.00	94.00	65
Tomato	320.00	480.00	50	30.00	18.40	39
Watermelon	240.00	450.00	88	33.00	21.00	36
Cotton	23.30	29.50	27	89.53	42.00	53
Lady finger	152.61	177.24	16	53.68	32.44	40
Egg plant	280.00	320.00	14	90.00	42.00	53
Bitter gourd	154.34	214.71	39	24.50	11.55	53
Ridge gourd	171.30	200.00	17	42.00	17.20	59
Cabbage	195.80	200.00	2	66.00	26.67	60
Papaya	13.40	23.48	75	228.00	73.30	68
Radish	70.45	71.86	2	46.41	10.81	77
Beet root	45.71	48.87	7	88.71	17.73	79
Chilly	42.33	60.88	44	109.71	41.77	62
Sweet potato	42.44	58.88	39	63.14	25.50	60

SOURCE: National Committee on the use of Plastics in Agriculture (NCPA), Status, potential and approach for Adoption of Drip and Sprinkler Irrigation System, Pune, 1990.

CHAPTER 4

WEED MANAGEMENT IN CROPS WITH MICRO IRRIGATION: A REVIEW

TIMOTHY COOLONG

CONTENTS

Modified from Coolong, Timothy, "Using Irrigation to Manage Weeds: A Focus on Drip Irrigation," Chapter 7, pp. 161–179. In *Weed and Pest Control: Conventional and New Challenges* by Sonia Soloneski and Marcelo Larramendy (Editors), ISBN 978-953-51-0984-6. http://www.intechopen.com/books/weed-and-pest-control-conventional-and-new-challenges/using-irrigation-to-manage-weeds-a-focus-on-drip-irrigation. Modified by the author. Originally published under Creative Commons Attribution 3.0 License.

4.1 INTRODUCTION

Irrigation has transformed agriculture and shaped civilization since its use in the Fertile Crescent more than 6000 years ago. Access to fresh water for irrigation transformed barren landscapes, allowing populations to move to previously uninhabitable regions. Advances in water management increased the productivity of agricultural systems around the world; supporting substantial population growth. Water consumption for agricultural use accounted for nearly 90% of global water use during the previous century [58] and is responsible for approximately 70% of fresh water withdrawals worldwide [60]. Currently, US water withdrawals for irrigation represent nearly 34% (137 billion gallons/day) of domestic water use [33]. Treating and pumping irrigation water has a significant carbon footprint as well. Pumping groundwater for irrigation requires about 150 kg Carbon/ha [38]. In the US more than 65% of total vegetable acreage and 76% of fruit acreage is irrigated [32]. Irrigating fruit and vegetable crops can increase marketable yields by 200% or more and is necessary to produce the high quality and yields required to be profitable [61]. It was estimated by Howell [32] that irrigated lands account for 18% of total cropped area, but result in approximately 50% of crop value.

Due to the large observable increases in yield and quality associated with irrigation, many growers over-irrigate, viewing it as an insurance policy for growing fruits and vegetables. Irrigation can routinely exceed 10% of input costs in the US [31] and over-irrigating may reduce yields in some instances [44]. Excessive irrigation not only depletes freshwater reserves, but may leach fertilizers and other chemicals from agricultural lands [14, 28, 67]. Unnecessary applications of water and fertilizer can also allow weeds to flourish. While irrigation systems are usually designed and managed with a crop of interest in mind; the impact of irrigation on weed growth is an important component of any modern production system.

This chapter will address the impacts of different irrigation systems on weed management with an emphasis on drip irrigation technologies [13].

4.2 THE IMPACT OF IRRIGATION METHODS ON WEED MANAGEMENT

Surface, sprinkler, and drip irrigation are the three primary types of irrigation methods used to grow crops (Fig. 4.1). Within each method, there are several subcategories, each of which varies in water use efficiency, cost, yield, and weed management potential.

FIGURE 4.1   Left: An irrigation canal for furrow irrigation of cabbage (*Brassica oleracea*); Center: Solid set sprinkler irrigation of onion (*Allium cepa*); and Right: Surface drip irrigation of recently planted cabbage.

4.2.1   IMPACT OF SURFACE IRRIGATION ON WEED MANAGEMENT

Surface irrigation, which floods entire fields or supplies water in furrows between planted rows, is the most common type of irrigation used worldwide. Some surface irrigation systems have been operating continuously for thousands of years and have the ability to supply enormous quantities of water over widespread areas. Flood and furrow irrigation can have water use efficiencies per unit of yield ranging from 25–50% of well managed drip irrigation systems [17]. One of the most common crops grown worldwide with flood irrigation is lowland rice (*Oryza sativa*). Flood irrigation can be an integral part of weed management for this crop.

As a semi-aquatic crop, lowland rice production uses substantial quantities of water. In studies on improving water use efficiency in rice, more than 10,000 m^3/ha of water was used to produce a crop according to typical agronomic practices [6]. This underscores the substantial water requirements for lowland rice production; particularly in the initial flooding stages when large quantities of water may be lost prior to saturation [70]. Although it has been reported that rice grown under saturated field conditions did not experience additional water stress and yielded no differently than rice grown under standing water [5, 65]; rice which is grown under standing water competes better with weeds than when grown in saturated soils [4, 45]. Although some weeds propagate vegetatively, most develop from seeds; thus flooding can restrict the germination and reduce the abundance of many weeds found in rice paddies [74].

Despite reducing the presence of some weed species, flooded lowland rice fields have selected for the presence of semi or aquatic weed species over time. To reduce the presence of some of these weeds, flooded soils are often tilled. While the primary goal of tillage is to uproot recently germinated weed seedlings; tilling flooded soils can destroy soil structure and porosity. This results in soils within low infiltration rates, which increases water retention, allowing fields to remain flooded [57].

Weed control in modern rice production is a system where irrigation management is integrated with tillage and planting practices as well as herbicides. Williams et al. [72] reported that weed control was improved in fields submerged under 20 cm of water compared to those submerged under 5 cm of water when no herbicides were used. However, when herbicides were included, weed control improved significantly at all depths [4]. Flowable granular herbicide formulations, which are often used in lowland rice production, also rely on standing water for dispersal. Flooded paddy fields allow uniform dispersal of low quantities of herbicides resulting in superior control of weeds [36, 71]. The integration of herbicides into the lowland rice production systems has reduced labor requirements for weed control by more than 80% since the introduction of 2,4-D in 1950, while simultaneously improving overall weed management [71]. Flooding has been an effective weed management technique in lowland rice for thousands of years. Coupled with modern herbicides, farmers can efficiently manage weeds on a large scale. Nonetheless, the high costs of water and demands on finite fresh water resources may result in substantial changes to the current lowland rice production system. The development of "aerobic-rice," drought tolerant lowland varieties that can yield well on nonsaturated soils, may change how irrigation is used to manage weeds in lowland rice. Aerobic-rice is grown in a manner similar to many other grains, with land allowed to dry between irrigation cycles. This has the potential to reduce the reliance on flooding and irrigation water for weed control, likely shifting to chemical or mechanical methods [69].

Furrow irrigation is a common irrigation method where water is sent through ditches dug between raised beds to provide water to plants. Instead of flooding entire fields, only furrows between beds are wetted, allowing water to seep into growing beds through capillary action. Furrow irrigation is commonly used on millions of hectares of crops worldwide; where complex canal networks can move irrigation water hundreds of miles from upland sources to lower elevation-growing areas. As would be expected, weed pressure in the irrigated furrows between rows is generally higher than with the rows themselves [27]. To control these weeds, mechanical cultivation may be used, but in many instances, herbicides, either applied to the soil as sprays or through irrigation water, are relied upon. The administration of herbicides through furrow irrigation can be challenging. Poor application uniformity, downstream pollution, and inaccuracies due to difficulties in measuring large quantities of water are challenges associated with applying herbicides through surface irrigation water [10, 48]. Chemical choice also is important when applying herbicides in surface irrigation systems. For example, Cliath et al. [10] noted that large quantities of the herbicide EPTC volatilized shortly after application via flood irrigation in alfalfa (*Medicago sativa*). Amador-Ramirez et al. [1] also reported variability in the effectiveness of some herbicides when applied through furrow irrigation compared to conventional irrigation methods.

A variant on the typical furrow irrigation system has been developed that combines furrow irrigation with polyethylene mulches and rainwater collection to irrigate crops, while controlling weeds. The production method, called the "ridge-furrow-ridge rainwater harvesting system," uses woven, water-permeable, polyethylene mulches that cover two ridges as well as a shallow furrow between the ridges [24, 42]. The system is similar to a raised-bed plastic mulch system, with inter-row areas being left in bare soil. However, unlike a traditional plastic mulch system, a furrow is made in the center of the raised bed to collect any rainwater that ordinarily would be lost as runoff from the bed. This system significantly reduces weed pressure in the furrow area and increases yield with the use of a polyethylene mulch, while reducing the need for supplemental irrigation by collecting rainwater [24]. Interestingly, a similar method of irrigation was employed during early experiments with plastic mulch, prior to the introduction of drip irrigation tubing. In these trials irrigation was achieved by cutting furrows in the soil next to the crop, covering them with plastic, and cutting holes in the plastic for the water to penetrate the plant bed [20, 21]

4.2.2 IMPACT OF SPRINKLER IRRIGATION ON WEED MANAGEMENT

Introduced on a large scale in the 1940 s, sprinkler irrigation systems are used on millions of ha of crop land. The three primary types of sprinkler irrigation are center pivot, solid set, and reel or traveling gun systems. Sprinkler systems require a pump to deliver water at high pressures and are costlier than surface irrigation systems, but provide superior application uniformity and require less water to operate [43, 55]. While center pivot systems require relatively level ground; solid set and reel-type systems can be used on with varied topographies. Because of improved application uniformity, sprinkler irrigation is the method of choice when applying herbicides or other agrichemicals through the irrigation system [26]. Sayed and Bedaiwy [56] noted a nearly 8-fold reduction in weed pressure when applying herbicides through sprinkler irrigation compared to traditional methods. Sprinkler irrigation permits growers to uniformly apply water over large areas, which can allow for proper incorporation of some pre-emergent herbicides [18]. In addition to applying herbicides, pre-plant sprinkler irrigation of fields, when combined with shallow tillage events after drying, has been shown to significantly reduce weed pressure during the growing season. This process of supplying water to weed seeds pri- or to planting, which causes them to germinate, where they can then be managed through shallow cultivation or through herbicide application is termed "stale seed-bedding" and is routinely used by farmers in many parts of the US.

4.2.3 IMPACT OF DRIP/TRICKLE OR MICRO IRRIGATION ON WEED MANAGEMENT

Introduced on a large scale in the late 1960s and early 1970s, drip irrigation has steadily grown in popularity [15]. Although drip irrigation is only used on approximately 7% of the total irrigated acreage in the US, it is widely used on high value crops such as berries and vegetables [33]. Drip irrigation, if properly managed, is highly efficient with up to 95% application efficiencies [53]. The productivity of drip irrigation has prompted significant increases (>500%) in its use over the previous 20–30 years [32]. While drip irrigation is typically expensive and requires significant labor to install and manage; the water savings compared to other methods of irrigation have prompted grower adoption. Drip irrigation has several benefits in addition to improved water use efficiencies. By only wetting the soil around plants, leaves are kept dry reducing foliar diseased and the potential for leaf burn when saline water is used for irrigation [15, 73]. Fertilizers, which are easily supplied through drip irrigation, are restricted to an area near the root zone. This leads to more efficient use by the target crop. Because drip irrigation wets the soil in the vicinity of the drip line or emitter, growers are able to supply irrigation water only in the areas required to grow the crop of interest. Soils between rows are not supplied with water or fertilizer, reducing weed growth. When drip irrigation is coupled with plastic mulch and preplant soil fumigation, weeds can be effectively controlled within rows, leaving only between-row areas to be managed. By restricting weed management to areas between rows growers increase their chemical and mechanical control options. While many farmers may apply preemergent herbicides to between-row areas, weeds that do germinate can be controlled easily with directed sprays of postemergent herbicides with low risk to the crops growing in the plastic mulch. In arid growing regions the combination of plastic mulch and drip irrigation may lead to acceptable weed control with- out the use of herbicides.

Because drip irrigation can supply limited quantities of water to an area immediately surrounding the crop root zone, it may be suitable for insecticide, fungicide, or nematicide injection. The small quantities of water delivered with drip irrigation requires significantly less chemical to maintain a given concentration applied to plants compared to surface or sprinkler irrigation [9]. However, many pesticides that are suited for injection in drip irrigation systems can be bound by soil particles. Therefore, distance from the target crop to the emitter is important. While drip irrigation is one of the most efficient means to deliver chemicals such as systemic insecticides to plants, it is much less effective than comparable sprinkler systems for herbicide applications. The limited wetting pattern and low volume of water used for drip irrigation means that herbicides do not reach much of the cropped area. Within wetted areas, herbicides may also be degraded prior to the end of the season [48]. Because drip systems are often designed for frequent, low-volume irrigations, soils around plants may remain moist, reducing the efficacy of preemergent herbicides. Fischer et al. [23] reported significantly better weed control when using micro sprinklers compared to drip irrigation in vineyards and orchards. This was due to a reduction in the effectiveness of preemergent herbicides in drip-irrigated treatments late in the growing season. The authors speculated that the drip irrigated plants had persistently greater soil moisture near the emitters resulting in enhanced degradation of the applied herbicides. Drip irrigation is often used in tandem with herbicides; however, they are often applied using conventional sprayers. Therefore, the weed control benefits of drip irrigation are due to the ability to precisely manage and locate water where it will most benefit crops while reducing availability for weed growth. One method that allows growers to precisely locate water in the root zone, below the soil surface, away from weed seeds is subsurface drip irrigation.

4.3   SUBSURFACE DRIP IRRIGATION

Subsurface drip irrigation (SDI) has been used in various forms for more than a century [8, 15, 25, 26]. Presently SDI uses standard drip irrigation tubing that is slightly modified for below-ground use. While typical surface drip irrigation tubing have walls that are usually 8 or 10-mil thick; tubing made specifically for multiseason SDI applications, have walls with a 15-mil thickness. In addition, tubing made specifically for SDI applications may have emitters, which are impregnated with herbicides to prevent root intrusion [75]. Because growers are unable to inspect buried tubing, any problems with emitter clogging or cuts in the line may go unnoticed for long periods of time. Subsurface drip irrigation used for the production of high-value crops such as vegetables, which tend to have shallow root systems, may be buried at depths of 15–25 cm [76]. Subsurface drip tubing that is used for agronomic crops such as cotton (*Gossypium* spp.) or corn (*Zea maize*) is generally buried 40–50 cm below the soil surface [40]. Drip irrigation tubing used for agronomic crops is typically left in place for several years in order to be profitable and must reside below the tillage zone to avoid being damaged [40]. Agronomic crops in general tend to be deeper rooted than many vegetable crops allowing them to access water supplied at greater depths. In addition, the deeper placement of the irrigation tubing reduces the potential rodent damage, which can be significant [40, 59].

Drip tubing may be placed during or after bed formation in tilled fields or into conservation tillage fields with drip tape injection sleds (Fig. 4.2). While SDI that is used for a single season may be connected to flexible "lay-flat" tubing at the ends of fields; more permanent installations are generally coupled to rigid PVC header lines.

Although concern over buried drip tubing collapsing under the pressure of the soil above is justified; properly maintained SDI systems have lasted 10–20 years in the Midwest US with- out significant problems [40]. For permanent systems, lines must be cleaned and flushed after every crop if not more frequently. In single-season trials conducted by the author, end of season flow rates were found to be no different between surface and SDI systems placed at a depth of 15 cm (T. Coolong, unpublished data). However, when comparing SDI that had been in use for three years for onion production to new SDI tubing, there were slight reductions in discharge uniformity in the used tapes [54].

FIGURE 4.2   Injection sled for SDI.

4.3.1 SUBSURFACE DRIP IRRIGATION IN ORGANIC FARMING

Some of the earliest uses of SDI were not based on enhanced water use efficiency (WUE) but because drip irrigation tubing on the soil surface could interfere with agricultural equipment, particularly cultivation tools [68]. While many conventional farmers now rely more on chemical weed control than on cultivation, most organic growers must rely exclusively on cultivation to manage weeds. For this reason, SDI is particularly appropriate for organic farming systems. Traditional placement of drip irrigation tubing requires growers to remove the tubing prior to cultivation, increasing labor costs. By burying drip tubing below the depth of cultivation, growers can control weeds mechanically. SDI is routinely used for bare-ground, organic vegetable production at The University of Kentucky Center for Horticulture Research (Lexington, KY, US). This system uses a SDI injection sled (Fig. 4.2) coupled with in-row cultivators to effectively control weeds in a humid environment (Fig. 4.3).

In this system, SDI tubing is placed approximately 15 cm below the surface on a shallow raised bed. Using SDI in combination with precision cultivation has allowed for nearly complete control of weeds on an organic farm in an environment which may regularly experience 25 cm or more rain during the growing season.

FIGURE 4.3   Crops that are grown with SDI and mechanical cultivation for near complete weed control in a humid environment. **Top left**: Buried drip irrigation tubing entering the soil at the end of a field; **Top right**: A two-row cultivator using side knives and spring hoes; **Bottom left**: A rolling basket weeder controlling weeds within and between rows; and **Bottom right:** Organically managed kale and collard (*Brassica oleracea Acephala* group).

4.3.2   SUBSURFACE DRIP IRRIGATION AND WATER USE

More than 40 types of crops have been tested under SDI regimes [8, 25, 26]. In most cases yields with SDI were no different than or exceeded yields for surface drip irrigation. In many cases water savings were substantial. However, SDI relies on capillary movement of water up-ward to plant roots. Soil hydraulic properties can significantly affect the distribution patterns of water around emitters, making interpretation of data difficult when comparing the effectiveness of SDI in different soil types [41]. Trials often report water savings or increased yield in SDI systems compared to surface drip systems for vegetable crop production [3, 39, 76], although some do not [59].

In 2012, studies were conducted at the University of Kentucky Center for Horti-culture Re- search (Lexington, KY, US) comparing SDI at a depth of 15 cm to sur-face placement of drip irrigation tubing for the production of acorn squash (*Cucur-bita pepo*) 'Table Queen.' The soil was a Maury silt loam. Irrigation was controlled automatically with switching – tensiometers placed at a depth of 15 cm from soil surface [11, 12]. Tensiometers were placed approximately 20 cm from plants and 15 cm from the drip tubing, which was centered on raised beds. Tensiometer set points were as follows: on/off −40/−10 kPa and −60/−10 kPa for both SDI and surface drip systems. In both moisture regimes the surface applied drip irrigation utilized less water during the growing season than SDI (Table 4.1). Interestingly, the number of irrigation events and the average duration of each event varied significantly among the surface and SDI treatments when irrigation was initiated at −40 kPa, but were similar when irrigation was scheduled at −60 kPa. Irrigations were frequent, but relatively short for the −40/−10 kPa surface irrigation treatment. Comparable re-sults have been reported in studies conducted in tomato (*Lycopersicon esculentum syn. Solanum lycopersicum*) and pepper (*Capsicum annuum*) using a similar man-agement system and set points. However, the SDI −40/−10 kPa treatment irrigated relatively infrequently and for longer periods of times. When irrigation was initiated at −60 kPa and terminated at −10 kPa there were differences in water use between the two drip systems, with the surface system being more efficient. However, un-like the −40/−10 kPa treatments, the numbers of irrigation events were not different between the two drip irrigation systems. The difference in the response of the SDI and surface systems when compared under different soil moisture regimes was not expected and suggests that irrigation scheduling as well as soil type may have a significant impact on the relative performance of SDI compared to surface drip ir-rigation. This should be noted when comparing the performance of SDI and surface drip irrigation systems.

4.3.3 SUBSURFACE DRIP IRRIGATION FOR IMPROVED WEED MANAGEMENT

As previously discussed, a key benefit of SDI is a reduction in soil surface wet-ting for weed germination and growth. Although the lack of surface wetting can negatively impact direct-seeded crops, transplanted crops often have significant root systems that may be wetted without bringing water to the soil surface. Direct-seeded crops grown with SDI are often germinated using overhead microsprinkler irrigation [39]. The placement of SDI tubing as well as irrigation regime [29] can impact the potential for surface wetting and weed growth. As mentioned previously, SDI is often located 40–50 cm below the soil surface in most agronomic crops, but is typically shallower (15–25 cm) for vegetable crops [39]. Patel and Rajput [49] evaluated five depths (0, 5, 10, 15, and 20 cm) of drip irrigation with three moisture regimes in potato (*Solanum tuberosum*). Soil water content at the surface of the soil was relatively moist for drip tubing placed 5 cm below the surface, while the soil surface remained relatively dry for the 10, 15, and 20 cm depths of drip tubing placement [49]. Because that study was carried out on sandy (69% sand) soils, greater depths may be required to prevent surface wetting on soils with a higher clay content and greater capillary movement of water [35].

TABLE 4.1   A Comparison of SDI and Surface Drip Irrigation Under Two Automated Irrigation Schedules

Irrigation treatment on/off	Irrigation type	Mean irrigation events	Mean irrigation time	Mean irrigation volume
kPa		no.	min/event	liters·ha–1
–40/–10	Surface	48	92	1.25 × 106
–40/–10	SDI	18	276	1.50 × 106
–60/–10	Surface	14	201	0.84 × 106
–60/–10	SDI	14	251	1.06 × 106

'Table Queen' squash grown with automated irrigation in 2012 in Lexington, KY.

SDI not only keeps the soil surface drier, but also encourages deeper root growth than surface drip systems. Phene et al. [51] reported greater root densities below 30 cm in sweet corn grown under SDI compared to traditional surface drip. In that study, the SDI tubing was placed at a depth of 45 cm. In bell pepper, a shallow rooted crop, SDI encouraged a greater proportion of roots at depths below 10 cm when laterals were buried at 20 cm [37]. Encouraging deeper root growth may afford greater drought tolerance in the event of irrigation restrictions during the production season.

In arid climates SDI has been shown to consistently reduce weed pressure in several crops, including cotton, corn, tomato, and pistachio (*Pistacia vera*) [16, 27, 64]. For example, weed growth in pistachio orchards in Iran was approximately four-fold higher in surface irrigated plots compared to those with SDI [16]. In humid regions, benefits may depend on the level of rainfall received during the growing season; however, a reduction in the consistent wetting of the soil surface should allow for a reduction in weed pressure, particularly when coupled with preemergent herbicides (Figure 4).

Processing tomatoes represent one of the most common applications of SDI in vegetable crops. The impact of SDI (25 cm below the soil surface) and furrow irrigation on weed growth were compared in tomato [27]. In that study the authors reported a significant decrease in weed growth in plant beds and furrows with SDI compared to furrow irrigation. When no herbicides were applied, annual weed biomass was approximately 1.75 and 0.05 tons per acre dry weight in the furrow and SDI treatments, respectively [27]. With herbicides, both irrigation treatments had similar levels of weed biomass. However, in that study, weed biomass in the SDI nonherbicide treatment was similar to the furrow irrigation with herbicide treatment, suggesting that when using SDI, herbicides may not be necessary in arid environments.

FIGURE 4.4   The difference in weed growth approximately 10 days after transplanting between acorn squash (*Cucurbita pepo*), which were subjected to SDI at a depth of 15 cm below the soil surface (**left**) and surface drip irrigation (**right**). A preemergent herbicide (halosulfuron methyl, Sandea™) was applied to all plots.

A similar trial compared SDI and furrow irrigation across different tillage regimes with and without the presence of herbicides in processing tomato [64]. In that study, both conservation tillage and SDI reduced the weed pressure compared to conventional alternatives. However, when main effects were tested, SDI had the largest impact on weed growth of any treatment. Main effects mean comparisons showed that SDI treatments had weed densities of 0.5 and 0.6 weeds per m2 in the planting bed in years one and two of the trial, respectively, compared to 17.9 and 98.6 weeds per m^2 in the plant bed for furrow irrigated treatments. As would be expected, SDI substantially reduced weed populations in the furrows between beds as they remained dry during the trial. In this trial SDI had a greater impact on weed populations than herbicide applications. The authors concluded that SDI could reduce weed populations sufficiently in conservation tillage tomato plantings in arid environments such that herbicides may not be necessary [64].

In another related trial, weed populations were evaluated for processing tomatoes grown with SDI and furrow irrigation under various weed-management and cultivation systems [59]. In that study, the authors noted an increase in weed densities in the furrow system compared to SDI within the planting bed and fur- rows.

However, there was no significant difference in the total weed biomass in the plant bed comparing the two irrigation systems [59]. The authors did note that the majority of the weeds in the SDI treatment were in the plant row and not evenly distributed across the bed, indicating that the outer regions of the plant bed were too dry to support weed germination or growth. Interestingly, when the relative percentages of weeds are extrapolated from the results provided, black nightshade, (*Solanum nigrum*) constituted 76% of the weed population in the plant beds of SDI treatments, but 52% of the weed population in the furrow irrigated beds. Although the sample size from that study is too small to make statements regarding selection pressures for weed species in the irrigation systems evaluated, it may give insight into why the authors reported a significant difference in numbers of weeds, but not biomass. *Solanum nigrum* can grow quite large and may have contributed a substantial amount of biomass in the SDI plots, despite having fewer numbers of weeds present. In this trial the furrow irrigation treatments had significantly greater yields than the SDI treatments [59]. The authors suggested that this was not due to a flaw in the SDI system, but poor management late in the season. The relatively small amounts of water used in drip irrigation underscore the need for proper scheduling; otherwise water deficits can occur, resulting in poor yields.

4.4 EFFICIENT MANAGEMENT OF DRIP/TRICKLE OR MICRO IRRIGATION

Appropriate management of irrigation requires growers to determine when and how long to irrigate. A properly designed and maintained drip irrigation system has much higher application efficiencies than comparable sprinkler or surface irrigation systems [15]. However, even with drip irrigation, vegetable crops can require large volumes of water – more than 200,000 gallons per acre for mixed vegetable operations in Central Kentucky, US [63]. Poorly managed drip irrigation systems have been shown to reduce yields [44] and waste significant quantities of water. Just 5 h after the initiation of drip irrigation, the wetting front under an emitter may reach 45 cm from the soil surface, effectively below the root zone of many vegetables [19]. When drip irrigation is mismanaged, a key benefit – limiting water available for weeds, is lost. The ability to precisely apply water with drip irrigation means that a very high level of management can be achieved with proper scheduling [42].

Irrigation scheduling has traditionally been weather or soil-based; although several plant-based scheduling methods have been proposed [22, 34]. In weather-based scheduling, the decision to irrigate relies on the soil-water balance. The water balance technique involves determining changes in soil moisture over time based on estimating evapotranspiration (Et) adjusted with a crop coefficient [50]. These methods take environmental variables such as air temperature, solar radiation, relative humidity and wind into account along with crop coefficients that are adjusted for growth stage and canopy coverage [22]. Irrigating based on Et can be very effective in large acreage, uniformly planted crops such as alfalfa, particularly when local weather data is available. However, irrigating based on crop Et values for the production of vegetable crops is prone to inaccuracies due to variations in microclimates and growing practices. Plastic mulches and variable plant spacing can significantly alter the accuracy of Et estimates [2, 7]. Furthermore the wide variability observed in the growth patterns in different cultivars of the same vegetable crop can substantially alter the value of crop coefficients at a particular growth stage. In many regions of the US, producers do not have access to sufficiently local weather data and the programs necessary to schedule irrigation.

An alternative to using the check-book or ET-based models for irrigation is to use soil moisture-based methods. Perhaps the simplest and most common method is the "feel method," where irrigation is initiated when the soil "feels" dry [46]. Experienced growers may become quite efficient when using this method. More sophisticated methods of scheduling irrigation may use a tensiometer or granular matrix type sensor [47, 52, 61, 66].

These methods require routine monitoring of sensor(s), with irrigation decisions made when soil moisture thresholds have been reached. This requires the development of threshold values for various crops and soil types. Soil water potential thresholds for vegetable crops such as tomato and pepper have been developed [61, 62, 66]. Drip irrigation is well suited to this type of management as it is able to frequently irrigate low volumes of water allowing growers to maintain soil moisture at a near constant level [6, 52, 53, 72]. In some soils, high-frequency, short-duration irrigation events can reduce water use while maintaining yields of tomato when compared to a traditionally scheduled high-volume, infrequent irrigation (Table 4.2) [12, 47].

Coolong et al. [12] reported that irrigation delivered frequently for short durations so as to maintain soil moisture levels in a relatively narrow range could save water and maintain yields, but efficiencies varied depending on season and the soil moisture levels that were maintained. In two years of trials, irrigation water was most efficiently applied when soil moisture was maintained between −45 and −40 kPa for tomatoes grown on a Maury Silt Loam soil. However, when soil moisture was maintained slightly wetter at −30 to −25 kPa, the relative application efficiency was affected by growing year (Table 4.2). Therefore, while an effective method, soil moisture-based irrigation scheduling may produce variable application efficiencies and should be used in concert with other methods.

TABLE 4.2 A Comparison of High Frequency Short Duration To More Traditional Infrequent But Long Duration Irrigation Scheduling Using Soil Moisture Tension To Schedule Irrigation [12]

Irrigation treatment on/off	Mean irrigation events	Mean irrigation time	Mean irrigation volume
kPa	no.	min/event	liters·ha^{-1}
2009			
−30/−10	39	110	1.30 × 106
−30/−25	59	91	1.63 × 106
−45/−10	21	221	1.41 × 106
−45/−40	76	40	1.92 × 106
2010			
−30/−10	28	144	1.22 × 106
−30/−25	22	140	0.93 × 106
−45/−10	22	167	1.11 × 106
−45/−40	18	146	0.79 × 106

After more than 40 years of research with drip irrigation, results suggest that a mix of scheduling tactics should be employed to most efficiently manage irrigation. The application efficiencies of several different management methods were determined by De Pascale et al. [17]. The authors estimated that when compared to a simple timed application, the use of soil moisture sensors to schedule irrigation would increase the relative efficiency of drip irrigation by 40–50%. Using a method incorporating climate factors and the water-balance technique, one could increase relative efficiency compared to the baseline by 60–70%. However, when soil moisture sensors were combined with Et-based methods, the relative efficiency of drip irrigation could be increased by more than 115% over a fixed interval method. Therefore, multiple strategies should be used to optimize drip irrigation scheduling. This ensures maintaining yields while reducing excessive applications of water, reducing the potential for weed growth.

4.5 SUMMARY

Irrigation management is essential to developing a holistic system for weed management in crops. As water resources become costlier, drip irrigation technologies will become more widely used by growers worldwide. Although drip irrigation may be adopted due to water savings, the impact of drip irrigation on weed control is noteworthy. The ability to reduce soil wetting will allow for improved weed control over sprinkler and surface irrigation systems. Furthermore, precisely locating water in the root-zone without wetting the soil surface will make SDI more attractive to

growers, despite the higher installation costs. In addition, SDI is now being implemented on large acreages for the production of grain crops, particularly corn, in
the Midwestern US. With the increase in adoption of SDI, new technologies will
be developed to overcome some of the limitations of that system. Future research
will likely continue to develop management tactics combining multiple scheduling
strategies such as ET and soil moisture-based irrigation [12] and its application for
managing SDI on a wider range of crops and soil types.

KEYWORDS

- aquatic crop
- cabbage
- center pivot
- crop coefficient
- domestic water use
- drip irrigation
- evapotranspiration, ET
- excessive irrigation
- flood irrigation
- fresh water
- fruit crops
- furrow irrigation
- groundwater
- irrigated lands
- irrigation
- irrigation canal
- irrigation scheduling
- irrigation water
- micro irrigation
- onion
- pepper

- pest control
- plastic mulch
- polyethylene mulch
- potato
- rice
- squash
- surface drip irrigation
- sweet corn
- tensiometer
- tillage
- tomato
- US
- vegetable crops
- water balance
- water management
- water requirement
- water stress
- water use efficiency
- weed control
- weed management
- wetting front

REFERENCES

1. Amador-Ramirez, M. D., Mojarro-Davila, F., Velasquez-Valle, R. (2007). Efficacy and economics of weed control for dry chili pepper. *Crop Protection*, 26(4), 677–682.
2. Amayreh, J., Al-Abed, N. (2005). Developing crop coefficients for field-grown tomato (*Lycopersicon esculentum Mill.*) under drip irrigation with black plastic mulch. *Agricultural Water Management*, 73(3), 247–254.
3. Beyaert, R. P., Roy, R. C., Ball-Coelho, B. R. (2007). Irrigation and fertilizer management effects on processing cucumber productivity and water use efficiency. *Canadian Journal of Plant Science*, 87(2), 355–363.
4. Bhagat, R. M., Bhuiyan, S. I., Moody, K. (1996). Water, tillage and weed interactions in lowland rice: a review. *Agricultural Water Management*, 31(3), 165–184.
5. Bhuiyan, S. I. (1982). *Irrigation system management research and selected methodological issues*. International Rice Research Institute Research Paper Series 81. Los Banos, Phillipines International Rice Research Institute.

6. Bhuiyan, S. I., Sattar, M. A., Khan, A. K. (1995). Improving water use efficiency in rice irriga-tion through wet seeding. *Irrigation Science*, 16(1), 1–8.

7. Burman, R. D., Nixon, P. R., Wright, J. L., Pruitt, W. O. (1980). Water requirements. In: *Jensen, M. E. (ed.) Design and Operation of Farm Irrigation Systems*. St. Joseph, Michigan: American Society of Agricultural Engineers. Pages 189–232.

8. Camp, C. R. (1998). Subsurface drip irrigation: a review. *Transactions of the ASAE*, 41(5), 1353–1367.

9. Clark, G. A., Smajstrla, A. G. (1996). Injecting chemicals into drip irrigation systems. *Hort-Technology*, 6(3), 160–165.

10. Cliath, M. M., Spencer, W. F., Farmer, W. J., Shoup, T. D., Grover, R. (1980). Volatilization of s-ethyl N, N-dipropylthiocarbamate from water and wet soil during and after flood irrigation of an alfalfa field. *Journal of Agricultural and Food Chemistry*, 28(3), 610–613.

11. Coolong, T., Snyde, r J., Warner, R., Strang, J., Surendran, S. (2012). The relationship between soil water potential, environmental factors and plant moisture status for poblano pepper grown using tensiometer – scheduled irrigation. *International Journal of Vegetable Science,* 18(2), 137–152.

12. Coolong, T., Surendran, S., Warner, R. (2011). Evaluation of irrigation threshold and duration for tomato grown in a silt loam soil. *HortTechnology*, 21(4), 466–473.

13. Coolong, Timothy, 2013. Using Irrigation to Manage Weeds: A Focus on Drip Irrigation. Chap-ter 7, pages 161–179. In: *Weed and Pest Control – Conventional and New Challenges* by S. Soloneski and M. Larramendy (ed.), ISBN 978-953-51-0984-6. http://www.intechopen.com.

14. Correll, D. L. (1998). The role of phosphorus in the eutrophication of receiving waters: a review. *Journal of Environmental Quality*, 27(2), 261–266.

15. Dasberg, S., Or, D. (1999). *Drip Irrigation*. Berlin: Springer.

16. Dastorani, M. T., Heshmati, M., Sadeghzadeh, M. A. (2010). Evaluation of the efficiency of surface and subsurface irrigation in dry land environments. *Irrigation and Drainage Eng.,* 59(2), 129–137.

17. De Pascale, S., Dalla Costa, L., Vallone, S., Barbieri, G., Maggio, A. (2011). Increasing water use efficiency in vegetable crop production: from plant to irrigation. *HortTechnology,* 21(3), 301–308

18. Dowler, C. C. (1995). Advantages of herbigation. *International Water and Irrigation Review,* 15(3), 26–29.

19. Elmaloglou, S., Diamantopoulos, E. (2007). Wetting front advance patterns and water losses by deep percolation under the root zone as influenced by pulsed drip irrigation. *Agricultural Water Management*, 90(1–2), 160–163.

20. Emmert, E. M. (1958). Black polyethylene for mulching vegetables. *Proceedings of the Amer-ican Society for Horticultural Science,* 69:464–469.

21. Emmert, E. M. (1956). Polyethylene mulch looks good for early vegetables. *Market Growers' Journal*, 85, 18–19.

22. Fereres, E., Golhamer, D. A., Parsons, L. R. (2004). Irrigation water management of horticul-tural crops. *HortScience,* 38(5), 1036–1042.

23. Fischer, B. B., Goldhamer, D. A., Babb, T., Kjelgren, R. (1985). Pre-emergence herbicide per-formance under drip and low volume sprinkler irrigation. In: Howell, T. A. (ed.) *Drip/Trickle Irrigation in Action Volume II*. Proceedings of the Third International Drip/ Trickle Irrigation Congress, 18–21 November, Fresno California. St. Joseph, Michigan: American Society of Agricultural Engineers.

24. Gosar, B., Baricevic, D. (2011). Ridge furrow ridge rainwater harvesting system with mulches and supplemental irrigation. *HortScience*, 46(1), 108–112.

25. Goyal, Megh R. (2014). *Book series: Research Advances in Sustainable Micro Irrigation*. Volumes 1 to 10. www.appleacademicpress.com. Oakville, ON, Canada: Apple Academic Press Inc.,

26. Goyal, Megh R. (2013). *Management of Drip/Trickle or Micro Irrigation*. www.appleacademicpress.com. Oakville, ON, Ca: Apple Academic Press Inc.

27. Grattan, S. R., Schwankl, L. J., Lanini, W. T. (1988). Weed control by subsurface drip irrigation. *California Agriculture*, 42(3), 22–24.

28. Hallberg, G. R. (1989). Pesticide pollution of groundwater in the humid United States. *Agriculture Ecosystems and Environment*, 26(3–4), 299–367.

29. Hanson, B., May, D. (2004). Effect of subsurface drip irrigation on processing tomato yield, water table depth, soil salinity, and profitability. *Agricultural Water Management*, 68:(1)1–17.

30. Hartz, T. K. (1996). Water management in drip-irrigated vegetable production. *HortTechnology*, 6(3), 165–167.

31. Hochmuth, G. J., Locascio, S. J., Crocker, T. E., Stanley, C. D., Clark, G. A., Parsons L. (1993). Impact of micro irrigation on Florida horticulture. *HortTechnology*, 3(2), 223–229.

32. Howell, T. A. (2001). Enhancing water use efficiency in irrigated agriculture. *Agronomy Journal*, 93(2), 281–289.

33. Hutson, S. S., Barber, N. L., Kenny, J. F., Linsey, K. S., Lumia, D. S., Maupin, M. A. (2004). *Estimated use of water in the United States in 2000*. Reston, Virginia: US Geological Survey.

34. Jones, H. G. (2004). Irrigation scheduling: advantages and pitfalls of plant-based methods. *Journal of Experimental Botany*, 55(407), 2427–2436.

35. Jury, W. A., Horton, R. (2006). *Soil Physics 6th Edition*. Hoboken: Wiley.

36. Kamoi, M., Noritake, K. (1996). Technical innovation in herbicide use. Japan International Research Center for Agricultural Sciences IRCAS International Symposium Series, 4:97–106.

37. Kong, Q., Li, G., Wang, Y., Huo, H. (2012). Bell pepper response to surface and subsurface drip irrigation under different fertigation levels. *Irrigation Science*, 30(3), 233–245.

38. Lal, R. (2004). Carbon emissions from farm operations. *Environment International*, 30(7), 981–990.

39. Lamm, F. R., Camp, C. R. (2007). Subsurface drip irrigation. In: Lamm F. R., Ayars J. E., Nakayama F. S. (eds.) *Micro irrigation for Crop Production*. Amsterdam: Elsevier. Pages 473–551.

40. Lamm, F. R., Trooien, T. P. (2003). Subsurface drip irrigation for corn production: a review of 10 years of research in Kansas. *Irrigation Science*, 22(3–4), 195–200.

41. Lazarovitch, N., Shani, U., Thompson, T. L., Warrick, A. W. (2006). Hydraulic properties affecting discharge uniformity of gravity-fed subsurface drip irrigation systems. *Journal of Irrigation and Drainage Engineering*, 132(6), 531–536.

42. Li, X. Y., Gong, L. D., Gao, Q. Z., Li, F. R. (2001). Incorporation of ridge and furrow method of rainfall harvesting with mulching for crop production under semiarid conditions. *Agricultural Water Management*, 50(3), 173–183.

43. Locascio, S. J. (2005). Management of irrigation for vegetables: past, present, and future. *HortTechnology*, 15(3), 482–485.

44. Locascio, S. J., Olson, S. M., Rhoads, F. M. (1989). Water quantity and time of N and K application for trickle irrigated tomatoes. *Journal of the American Society for Horticulture Science*, 114(2), 265–268.

45. Matsunaka, S. (1983). Evolution of rice weed control practices and research: world perspective. In: Swaminathan A. (ed.) *Weed Control in Rice*. Los Banos, Philippines: International Rice Research Institute. Pages 5–17.

46. Maynard, D. N., Hochmuth, G. J. (2007). *Knott's Handbook for Vegetable Growers* Fifth Edition. Hoboken: Wiley.

47. Munoz-Carpena, R., Dukes, M. D., Li, Y. C. C., Klassen, W. (2005). Field comparison of tensiometer and granular matrix sensor automatic drip irrigation on tomato. *HortTechnology*, 15(3), 584–590.
48. Ogg, A. G. (1986). Applying herbicides in irrigation water – a review. *Crop Protection*, 5(1), 53–65.
49. Patel, N., Rajput, T. B. S. (2007). Effect of drip tape placement depth and irrigation level on yield of potato. *Agricultural Water Management*, 88(1–3), 209–223.
50. Penman, H. L. (1948). Natural evaporation from open water, bare soil and grass. *Proceedings of the Royal Society of London Series A- Mathematical and Physical Sciences*, 193(1032), 120–145.
51. Phene, C. J., Davis, K. R., Hutmacher, R. B., Bar-Yosef, B., Meek, D. W., Misaki, J. (1991). Effect of high frequency surface and subsurface drip irrigation on root distribution of sweet corn. *Irrigation Science*, 12(3), 135–140.
52. Richards, L. A., Russell, M. B., Neal, O. R. (1938). Further developments on apparatus for field moisture studies. *Proceedings of the Soil Science Society of America*, 2:55–64.
53. Rogers, D. H., Lamm, F. R., Alam, M., Trooien, T. P., Clark, G. A., Barnes, P. L., Mankin, K. (1997). Efficiencies and water losses of irrigation systems. *Irrigation Management Series MF-2243*. Manhattan, Kansas: Kansas State University Cooperative Extension Service.
54. Safi, B., Neyshabouri, M. R., Nazemi, A. H. (2007). Water application uniformity of a subsurface drip irrigation system at various operating pressures and tape lengths. *Turkish Journal of Agriculture and Forestry*, 31(5), 275–285.
55. Sammis, T. W. (1980). Comparison of sprinkler, trickle, subsurface, and furrow irrigation methods for row crops. *Agronomy Journal*, 72(5), 701–704.
56. Sayed, M. A., Bedaiwy, M. N. A. (2011). Effect of controlled sprinkler chemigation on wheat crop in a sandy soil. *Soil and Water Research*, 6(2), 61–72.
57. Sharma, P. K., De Datta, S. K. (1986). Physical properties and processes of puddled rice soils. In: Stewart BA. (ed.) *Advances in Soil Science 5*. Berlin: Springer. Pages 139–178.
58. Shiklomanov, I. A. (2000). Appraisal and assessment of world water resources. *Water International*, 25(1), 11–32.
59. Shrestha, A., Mitchell, J. P., Lanini, W. T. (2007). Subsurface drip irrigation as a weed management tool for conventional and conservation tillage tomato (*Lycopersicon esculentum Mill.*) production in semiarid agroecosystems. *Journal of Sustainable Agriculture*, 31(2), 91–112.
60. Siebert, S., Burke, J., Faures, J. M., Frenken, K., Hoogeveen, J., Doll, P., Portmann, F. T. (2010). Groundwater use for irrigation – a global inventory. *Hydrology and Earth System Sciences Discussions*, 14:1863–1880.
61. Smajstrla, A. G., Locascio, S. J. (1996). Tensiometer – controlled, drip irrigation scheduling of tomato. *Applied Engineering*, 12(3), 315–319.
62. Smittle, Da., Dickens, W. L., Stansell, J. R. (1994). Irrigation regimes affect yield and water use by bell pepper. *Journal of the American Society for Horticultural Science*, 199(5), 936–939.
63. Spalding, D. (2008). On-farm commercial vegetable demonstrations. In Coolong T., Snyder J., Smigell S. (eds.) *2008 Fruit and Vegetable Research Report*. Lexington, Kentucky: University of Kentucky Cooperative Extension Service. Pages 11–12.
64. Sutton, K. F., Lanini, W. T., Mitchell, J. P., Miyao, E. M., Shrestha, A. (2006). Weed control, yield, and quality of processing tomato production under different irrigation, tillage, and herbicide systems. *Weed Technology*, 20(4), 831–838.
65. Tabbal, D. F., Lampayan, R. M., Bhuiyan, S. I. (1992). Water efficient irrigation technique for rice. In: *Proceedings of the International Workshop on Soil and Water Engineering for Paddy Field Management*, 28–30 January, Bangkok, Thailand. Asian Institute of Technology; Pathumthani, Thailand.

66. Thompson, R. B., Gallardo, M., Valdez, L. C., Fernandez, M. D. (2007). Using plant water status to define threshold values for irrigation management of vegetable crops using soil moisture sensors. *Agricultural Water Management*, 88(1–3), 147–158.

67. Tilman, D., Fargione, J., Wolff, B., D'Antionio, C., Dobson, A., Howarth, R., Schindler, D., Schlesinger, W. H., Simberloff, D., Swackhamer, D. (2001). Forecasting agriculturally driven global environmental change. *Science*, 292(5515), 281–284.

68. Tollefson, S. (1985). Subsurface drip irrigation of cotton and small grains. In: Howell TA. (ed.) *Drip/Trickle Irrigation in Action Volume II*: Proceedings of the Third International Drip/Trickle Irrigation Congress, 18–21 November (1985). Fresno California. St. Joseph, Michigan: American Society of Agricultural Engineers.

69. Tuong, T. P., Bouman, B. A. M. (2003). Rice production in water scarce environments. In: Kijne JW., Barker R., Molden D. (eds.) *Water productivity in agriculture: limits and opportunities for improvement*. Wallingford, UK: CABI. Pages 53–67.

70. Valera, A. (1976). Field studies on water use and duration for land preparation for lowland rice. MS Thesis. University of Philippines.

71. Watanabe, H. (2011). Development of low land weed management and weed succession in Japan. *Weed Biology and Management*, 11(4), 175–189.

72. Williams, J. F., Roberts, S. R., Hill, J. E., Scardaci, S. C., Tibbits, G. (1990). Managing water for weed control in rice. *California Agriculture*, 44(5), 7–10.

73. Yarwood, C. E. (1978). Water and the infection process. In: Kozlowsky TT. (ed.) *Water Deficits and Plant Growth, Volume 5*. New York: Academic Press. Pages 141–156.

74. Zimidahl, R. L., Moody, K., Lubigan, R. T., Castin, E. M. (1988). Patterns of weed emergence in tropical soil. *Weed Science*, 36(5), 603–608.

75. Zoldoske, D. F., Genito, S., Jorgensen, G. S. (1995). Subsurface drip irrigation (SDI) on turfgrass: a university experience. In: Lamm FR. (ed.) *Micro irrigation for a Changing World: Conserving Resources/Preserving the Environment*: Proceedings of the Fifth International Microirrigation Congress, 2–6 April, Orlando, Florida. St. Joseph, Michigan: American Society of Agricultural Engineers.

76. Zotarelli, L., Scholberg, J. M., Dukes, M. D., Munoz-Carpena, R., Icerman, J. (2009). Tomato yield, biomass accumulation, root distribution and irrigation water use efficiency on a sandy soil, as affected by nitrogen rate and irrigation scheduling. *Agricultural Water Management*, 96(1), 23–34.

CHAPTER 5

MICRO IRRIGATION TECHNOLOGY IN INDIA

N. ASOKARAJA

CONTENTS

5.1   INTRODUCTION

India with a geographical area of 329 million ha receives about 1170 mm rainfall with a runoff of 400 M ha-m. The surface flow is estimated to be 187 M ha-m. Out of total runoff, only 69 M ha-m is utilizable from surface water bodies and about 45 M ha-m runoff is estimated to be utilizable underground water resources. India needs 300 million tons of food grains to feed 1.6 billion people by 2050. To achieve the projected food grain production, Government of India is implementing various Centrally sponsored schemes like NMMI, NMSA, NHM, TANHODA, NCPAH, APMIP, GGRC projects to conserve water resources and improve the water use efficiency (WUE) to achieve *"More Crop Per Drop"* http://www.ncpahindia.com/mi.

Water is a renewable resource, and its availability in appropriate quality and quantity is under severe stress due to increasing demand from various sectors. Agriculture is the largest user of water, which consumes more than 80% of the country's exploitable water resources. The overall development of the agriculture sector and the intended growth rate in GDP is largely dependent on the judicious use of available water resources. While the irrigation projects (major and medium) have contributed to the development of water resources, the conventional methods of water conveyance and irrigation, being highly inefficient, has led not only to wastage of water but also to several ecological problems like water logging, salinization and soil degradation making productive agricultural lands unproductive. It has been recognized that use of modern irrigation methods like drip and sprinkler irrigation is the only alternative for efficient use of surface as well as ground water resources.

5.2   *STATUS OF DRIP AND SPRINKLER IRRIGATION IN INDIA*

Drip and sprinkler irrigation methods are being used in many developed countries like USA, Austria, Germany, Israel, Great Britain, etc. The area under drip and sprinkler irrigation is shown in Table 5.1. Micro irrigation is adopted in 100% of the total irrigated area in countries like Great Britain, Finland, Germany, Israel, Czech Rep, Austria, etc. The Government of India is promoting drip irrigation system by launching Centrally Sponsored Scheme on National Mission on Sustainable Agriculture (NMSA) and providing financial assistance up to 60% of project cost for small and marginal farmers and 50% for general farmers including 10% State share since 2005–2006. An area of 1.43 and 2.4.5 million ha has been covered under drip and sprinkler irrigation, respectively. Andhra Pradesh and Maharashtra are among the leading States in India covering nearly 33.7% of the area under drip irrigation followed by with 35.33%, Maharashtra with 24.26%, Gujarat with 13.6%, Karnataka with 9.19%, and Tamil Nadu with 4.41% of the total drip irrigated area in India [3].

TABLE 5.1   Drip and Sprinkler Irrigated Area – World Scenario (Area in 1000 ha)

Country	Total irrigated Area ×10³ ha	Area under drip & sprinkler ×10³ ha	Percentage of total irrigated area %
World	173277	27065	16
Great Britain	150	150	100
Finland	86	86	100
Germany	532	530	100
Israel	231	230	100
Czech Rep.	155	154	99
Austria	80	79	99
France	1575	1483	94
Italy	2535	1414	56
Spain	3315	1819	55
USA	25050	13145	52
South Africa	1300	475	37
Australia	2384	715	30
India	62200	55	5
China	53300	1467	3

SOURCE: www.icid.org, 2010.

TABLE 5.2   Selected State-Wise Area (In Hectares) Covered Under Drip and Sprinkler In India During 2008–2009

States	Drip	Sprinkler	Total
Andhra Pradesh	87381	36200	123581
Bihar	81.69	0	81.69
Chhattisgarh	2172.4	34121.56	36293.96
Delhi	0	0	0
Goa	5.39	70.71	76.1
Gujarat	34028	19399	53427
Haryana	2141.52	20160.17	22301.69
Jharkhand	0	0	0
Karnataka	22737.3	69885	92622.3
Kerala	947.85	580.93	1528.78
Madhya Pradesh	15971.5	22327.84	38299.34
Maharashtra	60011	34701	94712
Orissa	2100	582.53	2682.53
Punjab	2787.48	409.58	3197.06
Rajasthan	5097	72632	77729
Tamil Nadu	10906.1	667.71	11573.81
Uttar Pradesh	921.48	366	1287.48
West Bengal	55.6	0	55.6
Total	**247345.3**	**312104**	**559449.3**

SOURCE: Rajya Sabha (upper house of the Parliament of India) unstarred question No.731, dated on 10.07.2009 [3].

TABLE 5.3 State-Wise Area (In Hectares) Covered Under Drip and Sprinkler Irrigation System In India (2010–2011 and 2011–2012 (Upto January 2012)) [3]

States	2010–11	% of total	2011–12 (till Jan., 2012)
Andhra Pradesh	122758	17.79	91774
Bihar	13485.04		14620.80
Chhattisgarh	21830.93		16129
Goa	119.065		34.00
Gujarat	78294	11.34	60492
Haryana	9340.2		2556.92
Jharkhand	1217.1		0.00
Karnataka	87447	12.67	36695
Kerala	2340.01		3078.64
Madhya Pradesh	41238.24		36544.88
Maharashtra	118025.08	17.10	70116.86
Odisha	12013.96		8605.24
Punjab	4925		4026.31
Rajasthan	147613	21.39	87207
Tamil Nadu	26153.16	3.79	14228.05
Uttar Pradesh	3108.63		3419.86
West Bengal	294		0
Arunachal Pradesh	0		0
Mizoram	0		0
Meghalaya	0		0
Tripura	0		0
Sikkim	0		0
India	**690202.42**		**449528.56**

SOURCE: Lok Sabha (lower house of the Parliament of India) unstarred question No. 1044, dated on 20.03.2012 [3].

However, according to Table 5.3, Rajasthan state is the leading State in India with highest percentage of area under micro irrigation (mostly under sprinkler) followed by Andhra Pradesh (17.78%), Maharashtra (17.10%), Karnataka (12.67%), Gujarat (11.34%) and Tamil Nadu with only 3.79%. Drip irrigation is suitable for all row crops and especially for wide spaced high value crops. The required quantity of water is provided to each plant at the root zone through a pipe network. Hence there is little loss of water from the soil surface. Micro irrigation is well adapted for undulating terrain, shallow soils, porous soils and water scarce areas.

Micro irrigation, which includes drip and micro sprinklers, is an effective tool for conserving water resources and studies have revealed significant water saving ranging between 40 and 70% by drip irrigation compared with surface irrigation, with yield increases as high as 100% in some crops in specific locations (Table 5.4). Micro irrigation is very popular in wide spaced horticultural crops like coconut, mango, guava, sapota, pomegranate, lime, oranges, grapes, banana, tapioca, turmeric and close spaced crops like vegetables, potato and flowers like jasmine, rose, etc.

TABLE 5.4   Water Use and Crop Yield Under Micro and Conventional Irrigation Methods [5]

Crop	Irrigation method	Water requirement	% water saving	Yield	% increase in yield	Water use efficiency
		cm	%	Kg/ha	%	Kg/(ha-mm)
Banana	Drip	97.00	45.00	87500	52.00	90.20
	Surface	176.00	–	57500	–	32.67
Sugarcane	Drip	94.00	56.00	170000	33.00	180.85
	Furrow	215.00	–	128000	–	59.53
Grapes	Drip	27.80	48.00	32500	23.00	116.90
	Surface	53.20	–	26400	–	49.62
Cotton	Drip	28.00	66.27	3250	25.00	116.10
	Furrow	83.00	–	2600	–	31.33
Sugar beet	Drip	37.10	25.05	48990	17.09	1320.00
	Furrow	49.50	–	41840	–	850.00
Sweet pepper	Drip	48.00	2.04	11952	0.80	249.00
	Furrow	49.00	–	11858	–	242.00
Sweet potato	Drip	25.20	60.06	5888	38.73	233.65
	Surface	63.10	–	4244	–	67.26
Beetroot	Drip	17.70	79.34	887	55.34	50.11
	Surface	85.70	–	571	–	6.66
Radish	Drip	10.80	75.72	1186	13.49	109.80
	Surface	46.40	–	1045	–	22.52
Papaya	Drip	73.88	67.89	23490	69.47	0.32
	Surface	225.80	–	13860	–	0.06
Mulberry	Drip	20.00	60.00	71400	3.03	3570
	Surface	50.00	–	69300	–	1386
Tomato	Drip	18.40	39.00	48000	50.00	260.86
	Surface	30.00	–	32000	–	106.66

SOURCE: WTC Annual Reports 1985–2003, [5].

5.3   TYPES OF DRIP IRRIGATION SYSTEM

5.3.1   BASED ON THE INSTALLATION LOCATION OF DRIP LATERALS

5.3.1.1   SURFACE DRIP IRRIGATION

It is the system in which drippers and laterals are laid on the soil surface. The commonly used drippers in this system are nonpressure-compensating, pressure compensating drippers, inline drippers, adjustable discharge drippers and micro-tubings. The choice of these drippers depends on the type of crop, topography and soil type.

5.3.1.2   SUBSURFACE DRIP IRRIGATION

In this system water is applied slowly below the soil surface through drippers. This includes twin wall type systems. The most commonly used systems are bi-wall, turbo tape and typhone systems. Subsurface system is mostly used for row crops.

5.3.2   BASED ON THE TYPE OF DRIP LATERALS AND/OR THE EMITTING DEVICES

5.3.2.1   OVERHEAD SYSTEM

Here the laterals taken off from sub main are laid overhead and drippers drip water from above the soil surface. This system is mostly used in grape fields (vine yards), which also produces conducive microclimate.

5.3.2.2   LINE SOURCE EMITTERS

Line source drip systems are generally used for row and vegetable crops such as squash, melons, asparagus, tomatoes, onions and peppers. More durable subsurface drip lines and above ground retrievable hoses are now available.

5.3.2.3   POINT SOURCE EMITTERS

Small fruits like strawberries, blueberries, blackberries, juneberries and raspberries respond well to microirrigation. The point source mode is suited to wider-spaced plants such as fruit trees and in vineyard.

5.3.2.4   ON LINE DRIP IRRIGATION SYSTEM

Drippers or Emitters are fixed on the Lateral Pipes by punching suitable holes on the drip lateral pipes at the locations specific to the crop being irrigated. These drippers are hence also called point source drippers. For trees and orchards, On-Line emitters are used so that only the root zone beneath them is watered.

5.3.2.5   IN LINE DRIP IRRIGATION SYSTEM

Drippers are factory installed within or on the drip lateral at regular intervals and are suitable for closely spaced field crops in order to achieve a continuous strip of wetting along the crop rows.

5.3.2.6   SPRAYER/MISTER/FOGGER/MINI SPRINKLER TYPE IRRIGATION SYSTEMS

These consist of emitting devices, which spray or sprinkle the water under pressure and wet the area around them in the diameter ranges of 1 to 5 m. The unit application rates are higher and the diameter of coverage is lower than that of the sprinkler type irrigation systems. These are useful for under tree irrigation of large trees, whose root zones are extensive, for completing irrigation in a shorter period of time, for ir-rigating/ misting/fogging applications in Nurseries, Shade houses, Greenhouses and other forms of controlled environment agriculture.

FIGURE 5.1   Non-pressure compensating dripper.

FIGURE 5.2   Online pressure compensating emitters.

FIGURE 5.3   Inline drippers or inline PE tubes.

FIGURE 5.4   Micro jet.

FIGURE 5.5   Mini sprinkler.

FIGURE 5.6   Fogger.

FIGURE 5.7   Bubbbler.

5.4   TYPES OF DRIPPERS OR EMITTERS

Non-pressure compensating dripper or pressure compensating drippers, in-line drippers, adjustable discharge type drippers, vortex type drippers and micro tubing of 1 to 4 mm diameter (Fig. 5.1). The choice among these drippers depends on the pressure discharge relationship of the dripper, the type of crop, topography and soil type of the area. In this system, water is applied to the soil near the root zone of the plants.

Online pressure-compensating devices (Fig. 5.2) are able to deliver the correct flow rate over a fairly wide range of inlet pressures, and within that range their flow rates are relatively constant. With the help of these drippers, water can be applied uniformly on long rows and on uneven slopes. A flexible internal diaphragm or disc inside the emitter changes shape at higher pressures to create greater restriction to flow as the pressure rises.

Online nonpressure compensating dripper does not possess the pressure compensation feature, but installed on the laterals; and hence the discharge tends to vary with operating pressure much widely. These drippers will be ideally suitable for the fields with little or no undulations and almost flat terrains. These drippers are less costly and are most widely used.

Inline drippers or inline PE tubes (Fig. 5.3): In row crops, a continuous wetting pattern is essential for irrigating all the plants in the rows. For achieving such continuous wetting pattern, the spacing between the drippers will be as close as 30 to 100 cm. In order to reduce the labor cost for installing drip systems in row crops and on large areas, inline PE tubes are suitable. In these inline tubes, drippers are inserted into the tube at the time of extruding the lateral tubes in the factory. The drippers are inserted at desired intervals based on the crop and soil requirements. They have very high emission uniformity of 94% per 300 m length of the lateral. The discharge of integral drippers ranges from 1.0–3.0 lph. They are clogging resistant, self-cleaning and uniform discharge. It is easy to roll and unroll the laterals. They operate at 0.5 to 3.0 bars/cm^2 of pressure. They are very much suitable to irrigate vineyards, plantations, orchards, row crops, cotton, maize and tomato.

Micro tubing: The technique involves the application of water through small tubes of internal diameters ranging from 0.5 to 1.2 mm. The longer the micro tube, the lower will be the discharge from it. The discharge of micro tubes varies from 1 lph for a length of 6' micro tube to 10 lph from a micro tube of length 6 inches.

Micro Jets or micro spray (Fig. 5.4): The main advantages of the micro jets or micro sprays are low operating pressure requirements of 0.8 to 1.5 kg/cm^2, low water application rates, and suitability for stony or very coarse sandy soils. These can meet the requirements of under-tree frost protection as well. In windy conditions, the jets/sprays with low angle of emission need to be installed.

Micro and mini sprinklers (Fig. 5.5): The spinners rotate with water pressure and sprinkle the water. Due to the impact created by the rotating spinners, the water can be sprinkled on a larger area than jets/sprays. These are useful for high discharge requirements in case of orchards and also where the crop canopy as well as the root zone spread is wider. They are similar to sprinklers but with less discharge rate. The spraying diameter varies from 3 to 10 m, and at 1 to 2 kg/cm^2 of operating pressure. It delivers 20–200 lph. The spraying diameter can be adjusted to 180° or 360°.

Flow static sprayer: It is otherwise called as Tornado or micro jet sprayer. It has no moving parts. It is available at various wetting diameters like 180°, 300° 360°. It is also available at various discharge rates of 25, 34, 55 and 70 lph at 1.5 bars of pressure. The wetting diameter ranges from 2.4 to 4.0 m. It is very much suited for sandy soils where the direct and excess application of water leads to percolation.

Foggers/Misters (Fig. 5.6): The foggers and misters are designed to produce very fine droplets creating a uniform cloud of miniature droplets, for applications such as: cooling, humidification and misting as well as for irrigating plants in greenhouses and nurseries. The fine mist and fog creates suitable microclimate necessary for germination and plant propagation requirements.

Bubbler systems (Fig. 5.7): These are low-pressure emitters designed to bubble-out water with rapid discharge rates. The bubbling patterns will be either in the form of sheet of water like an umbrella or in the form of streams. The typical flow rate from bubbler is between 8 and 80 lph. The irrigation water delivered by each bubbler is distributed uniformly by filling small basins, surrounded by low ridges. These are suitable in situations where large amounts of water need to be applied in a short period of time and suitable for irrigating trees with wide root zones and high water requirements. Despite many advantages, bubbler systems have not been widely used.

Drip tape: The drippers are fixed in a PE tubing at desired intervals in the factory. They are available at various dripper spacing's depending upon the crop and soil. They have the discharge rate of 2 to 4 L/meter/hour (0.2 to 0.4 lph/dripper). This system is ideally suited for vegetable and fruit crops. QueenGil drip tape was introduced in Israel in 1999.

Microtal drippers: They have very low flow rates suited for the crops growing in inert media, small containers and are suitable for light or sandy soils. The discharge is only 150–500 mL/dripper/hour. The main dripper discharge is 4 lph. From the main dripper, 16 microtal drippers (spaghettis) are connected. It is suitable for potted plants and for sloping lands.

Self compensated katif drippers: They are button like drippers, having self-cleaning mechanism. The discharge rate is from 2.0 to 4.0 lph. Since the drippers are very small, the rolling and unrolling of laterals becomes easy. These drippers are suitable for sandy soils and for vegetables, vineyards, row crops and orchard crops.

Low capacity sprinklers: These have a discharge rate of 80–250 lph, at 1.0 to 3.0 bar/cm² of pressure. The spacing between two sprinklers can be up to 7 m. It sprays water into very fine drops that is needed for sensitive crop germination.

Automatic irrigation controller (Irrimaster): The Irrimaster controller is the computer-based irrigation controlling system and it irrigates field by quantity and time. One controller unit can control up to 6 valves. If there is any blockage or rupture in the line it will automatically stops irrigation and send signals to the operator.

Pop-up sprinkler: The pop up sprinkler is used to irrigate lawns. The sprinkler head is above the ground while irrigating, and after irrigation the head is automatically goes inside the ground. So the moving operation can be done without any obstacle. Pop-up sprinklers have discharge of 0.45 to 1.18 m³/hr. The spacing between two units can be extended up to 12 m. It is resistant to clogging and can cover a radius of 15–40 feet.

Rain gun: It is a novel irrigation system to irrigate larger area at a time. It has a discharge up to 19 lps (liters per second). The discharge can be adjusted depending on the crop water requirement, soil type and operating pressure. It has an adjustable rotation to cover 90, 180, 270 and 360° angle coverage. This system is suitable for sugarcane, vegetables, fodder crops, cotton, wheat, groundnut, coffee, tea, turmeric and tobacco. It operates with 3 HP motor.

5.5   AUTOMATIC SYSTEM

The automatic filtration system consists of a disc filtration system for irrigating larger area, for a discharge rate up to 4000 lph. The discharge is continuous and steady. It operates automatically. If there is a pressure decrease or reduced discharge, it automatically stops filtering and starts back flushing to remove the particles that cause blockage and then it starts filtration again. These filtration systems have wider applications and multiple uses like industry, filtering of drinking water in large distribution systems and filters water for agricultural irrigation.

Automation Equipment: The micro irrigation systems are designed to be operated in a given sequence of opening and closing of valves as per the irrigation schedules. However, in order to automate the irrigation scheduling, electronic programmable controllers (Fig. 5.8) as well as host of sensors (Fig. 5.9) along with Irrigation Management software, are available. Use of such automation equipment further enhances the efficiency of the drip irrigation systems, as water and fertilizers can be applied with better control.

FIGURE 5.8   Time based irrigation controller.

FIGURE 5.9   Tensiometer based irrigation scheduling.

FIGURE 5.10   An aerial view of
sprinkler irrigation.

FIGURE 5.11   Sprinkler head.

There are two types of automation systems: (i) Time based automation; and (ii) Sensor based automation. In time based automation systems, the operation of valves is based on prefixed irrigation start times and end times and the sequence of valve operations required. In sensor based automation systems, sensors either individually or in combination can be attached to the controller for activating and deactivating the low voltage electric solenoid valves and pumps so that the irrigation system can be operated automatically based on the sensor readings.

5.6   SPRINKLER IRRIGATION

A sprinkler "throws" water through the air in an effort to simulate rainfall, whereas the other irrigation methods apply water directly to the soil or root zone, either on or below the surface. When many sprinklers are used they are attached to a pipeline at a predetermined spacing in order to achieve a uniform water application (Figs. 5.10 and 5.11). There are two major types of sprinkler systems on the basis of the arrangement for spraying or sprinkling water: Rotating head or revolving sprinkler system; and Perforated pipe system. Rotating head or revolving sprinkler system can be divided into four categories namely:

FIGURE 5.12   Center pivot irrigation
system.

FIGURE 5.13   Rotating gun.

1 CONVENTIONAL ROTARY SPRINKLER/CONVENTIONAL SPRINKLER IRRIGATION SYSTEM

The sprinklers operate at low to medium pressure of 2 to 4 kg/cm^2 and can wet an area of 9 to 24 m wide and up to 300 m long at one setting. Application rates vary from 5 to 35 mm per hour.

2 BOOM TYPE/SELF-PROPELLED SPRINKLER SYSTEM

This system employs one boom sprinkler on each lateral. Boom sprinklers are moved by towing the towers to the next position along the laterals with a tractor or winch. The large sprinkler irrigates a width of 75 to 100 m depending on nozzle sizes and pressure and is particularly useful for tall crops such as maize and sugar cane, where space at regular intervals is available for maneuvering the portable towers.

3 CENTRE PIVOT SYSTEM

This self-propelled sprinkler system rotates around the pivot point and has the lowest labor requirements of all the sprinkler systems (Fig. 5.12). It is constructed using a span of pipes connected to moveable towers. It can irrigate approximately 130 acres with a square quarter section. Sprinkler packages are available for low to high operating pressures (2.0 to 6 kg/cm^2 at the pivot point). Center pivots are adaptable for any crop height and are particularly suited to lighter soils. They are generally not recommended for heavy soils with low infiltration rates.

4 RAIN GUN/LARGE ROTARY SPRINKLERS (FIG. 5.13)

This system operates at high pressure to irrigate large areas. They can irrigate areas up to 4 ha at one setting with an application rate varying from 5 to 35 mm per hour. There are two main types of the system: (i) hose pull system; and (ii) hose reel system.

5.7 SUMMARY

Surface irrigation will still dominate as the primary irrigation method in India. However, with the current trends, the area under micro irrigation will continue to expand [1, 2, 4]. In some countries, micro irrigation is adopted in 100% of the total irrigated area. Similarly in India, micro irrigation is being promoted by Central and State Government schemes like NMMI, NCPAH, NHM, TANHODA, NAPD, IAMWARM projects with World Bank funding support as well. States like Andhra Pradesh and Gujarat have recently expanded the area under micro irrigation through

APMIP and Kuppam Projects with GGRC (Gujarat Green Revolution Company), respectively and occupy leading positions in micro irrigated area in India.

Recently micro irrigation is popular among closely spaced crops due to water and labor scarcity. Various newer irrigation products are being introduced by irrigation industries for various applications. Online drip irrigation system is slowly being replaced with inline drip system owing to convenience in lay and relay of drip laterals. Similarly micro sprinklers are getting popular particularly in groundnut, curry leaf and leafy vegetables besides in fruit crops. Adoption of micro irrigation can save irrigation water to a larger extent. Recently micro jet, micro sprinkler, mini sprinklers, foggers, misters, humidifiers etc. are being widely used under greenhouse conditions. Subsurface drip irrigation with fertigation is widely adopted in Tamil Nadu especially in sugarcane. Based on the studies conducted on different horticultural crops, it has been found that adoption of this technology improves the yield and quality of crops. It is also highly beneficial to farming community to reduce the cost of production.

Although micro irrigation is an accepted technology by the Indian farmers, yet its adoption is very slow mainly due to high initial costs in horticultural crops, which varies from Rs. 40,000 to Rs. 95,000 per hectare. Government of India has at present extended subsidies for varying categories of farmers, up to 100% of the micro irrigation system costs. Hence, there is a wide scope for adoption of this technology especially for high value crops like flowers, fruits, vegetables and horticultural plantations as well as sugarcane, cotton, hybrid maize would have shorter pay-back periods.

KEYWORDS

- commercial crops
- cotton
- drip irrigation
- horticultural crops
- hybrid maize
- micro sprinklers
- sprinkler irrigation
- subsurface drip irrigation, SDI
- sugarcane
- surface drip irrigation
- Water Technology Center, WTC

REFERENCES

1. ICID, (2002). www.icid.org. International Committee on Irrigation and Drainage.
2. ICID, (2010). www.icid.org. Annual Report of International Committee on Irrigation and Drainage, 2009–2010
3. indiastat.com, (2008). Selected state-wise area covered under micro irrigation in India (As on 31.10.2008)
4. Reinders, F. B. (2000). Micro irrigation: A world over view. Paper presented in the 6th International Micro Irrigation Congress "Micro Irrigation Technology for Developing Agriculture," held at South Africa 22–27, Oct. 2010.
5. WTC, (1985–2003). Annual reports of Water Technology Center, Tamil Nadu Agricultural University, Coimbatore.

CHAPTER 6

ADVANCES IN FERTIGATION FOR MICRO IRRIGATION: INDIA

NATARAJAN ASOKARAJA

CONTENTS

6.1 PRINCIPLES OF FERTIGATION TECHNOLOGY

An approach for efficient utilization of water and fertilizers is necessary for agriculture in the twenty-first century. In other words, fertigation is addition of fertilizers to irrigation water and application via drip or similar micro irrigation system [29]. Fertigation provides Nitrogen, Phosphorous and Potassium as well as the essential trace elements (Mg, Fe, Zn, Cu, Mo, Mn) directly to the active root zone, thus minimizing the loss of expensive nutrients, which ultimately helps in improving productivity and quality of farm produce. Fertigation is by far the most common, and in some cases the only method of fertilizing the green houses, orchard, vegetables and drip irrigated field crops such as cotton, maize etc. The characteristics of fertilizers suitable for fertigation [6, 13] are listed below:

- high nutrient content in the solution;
- fully soluble at field temperature;
- fast dissolution in irrigation water;
- fine grade, flowable;
- no clogging of filters and emitters;
- low content of insoluble salts;
- minimum content of conditioning agents;
- compatible with other fertilizers;
- minimal interaction with irrigation water;
- no drastic changes of water pH (3.5<pH<9);
- low corrosivity of control head, irrigation system, and accessories.

Nitrogen is the nutrient mostly commonly used in fertigation with MIS, flood irrigation, moving laterals and overhead sprinkling systems. In general, all N fertilizers cause few clogging and precipitation problems with the exception of Ammonium Sulphate, which may cause precipitation of CaSO in hard, calcium-rich water. Urea is well suited for injection in micro irrigation system. It is highly soluble and dissolves in nonionic form so that it does not react with other substances in the water. Also urea does not cause precipitation problems.

Application of Phosphorous to irrigation water may cause precipitation of phosphate salts. The precipitation of insoluble di-calcium phosphate and di-magnesium phosphate and Fe-P compound in irrigation pipes and water emitters is likely in water with a high pH, and low pH, respectively. Reducing the pH of irrigation water will significantly reduce the risk of Ca-P compounds precipitation. Thus Phosphoric acid appears to be more suitable for fertigation.

Application of K fertilizers does not cause any precipitation of salts, except when using K SO4 with irrigation water containing high concentrations of Ca. Potassium ion is adsorbed at the cation exchange sites of soil colloids, but researchers have shown lateral and downward mobility of potassium when applied via drip irrigation. Research studies have indicated that the distribution of potassium is more uniform than that of either nitrate or phosphate.

Potassium Chloride for fertigation: Among red and white color Potash, white (fertigation grade) is the preferred form for fertigation due to its high solubility compared to red potash even though both types contain 60% K O. White potash gives a clear, clean and pure solution, while red potash solution contains iron impurities, which can clog the drippers. Potash is compatible with N and P fertilizers, besides having high K content in the irrigation solution.

Fertigation can increase the fertilizer use efficiency (Table 6.1). Tables 6.2–6.4 indicate fertilizers for fertigation of N, P and K through irrigation systems. Irrigation water, in which fertilizers are to be dissolved, should have pH levels between 5.8 to 7.8. Most of the specialty water soluble fertilizers are imported in India and marketed by dealers of irrigation systems and fertilizers (Table 6.5).

TABLE 6.1 Fertilizer Efficiencies For Various Application Methods [21]

Nutrient	Fertilizer use efficiency (FUE, %)	
	Soil application	Fertigation
Nitrogen	30–50	95
Phosphorous	20	45
Potassium	50	80

TABLE 6.2 N Fertilizers Used in Fertigation

Fertilizers	Grade	Formula	pH (1 g/L at 20 °C)
Ammonium nitrate	34 – 0 – 0	NH NO	
Ammonium sulfate	21 – 0 – 0	(NH) SO	7.0
Calcium nitrate	15 – 0 – 0	Ca(NO)	8.0
Di ammonium phosphate	21 – 53 – 0	(NH) HPO	4.9
Magnesium nitrate	11 – 0 – 0	Mg(NO)	5.8
Mono ammonium phosphate	12 – 61 – 0	NH H PO	5.7
Potassium nitrate	13 – 0 – 46	KNO	5.8
Urea	46 – 0 – 0	CO(NH)	
Urea ammonium nitrate	32 – 0 – 0	CO(NH). NH NO	5.5

TABLE 6.3 P Fertilizers Used in Fertigation

Fertilizers	Grade	Formula	pH (1 g/L at 20 °C)
Diammonium phosphate	21 – 53 – 0	(NH) HPO	8.0
Mono ammonium phosphate	12 – 61 – 0	NH H PO	4.9
Monopotassium phosphate	0 – 52 – 34	KH PO	5.5
Phosphoric acid	0 – 52 – 0	H PO	2.6

TABLE 6.4 K Fertilizers Used in Fertigation

Fertilizers	Grade	Formula	pH (1 g/L at 20 °C)z	Other nutrients
Monopotassium phosphate	0 – 52 – 34	KHPO		52% PO
Potassium chloride č	0 – 0 – 60	KCl		46% Cl
Potassium nitrate	13 – 0 – 46	KNO		13% N
Potassium sulfate	0 – 0 – 50	KSO		18% S
Potassium thiosulfateß	0 – 0 – 25	KSO		17% S

TABLE 6.5 Specialty Water Soluble Fertilizers Available in Market

Name	N	PO	KO	Micro nutrients*
Polyfeed (All 19)	19	19	19	1000 ppm Fe
Polyfeed (All 20)	20	20	20	500 ppm Mn
Polyfeed	11	42	11	75 ppm Zn
Polyfeed	16	8	24	200 ppm B
Polyfeed	19	19	19	35 ppm Mo
Polyfeed	15	15	30	55 ppm Cu
Mono Ammonium Phosphate (MAP)	12	61	0	
Potassium Nitrate (Multi-K)	13	0	46	
Mono Potassium Phosphate, MKP	0	52	34	
Sulphate of Potash, SOP	0	0	50	

* All Poly feed fertilizers contain all of the 6 micro nutrients
(Fe, Mn, Zn, B, Mo, Cu) at specified concentration.

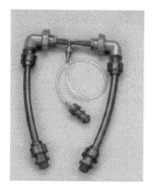

FIGURE 6.1   Venturi injector.

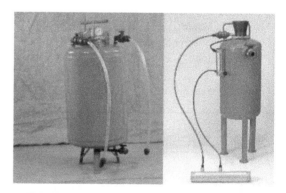

FIGURE 6.2   Fertilizer tank based bypass system.

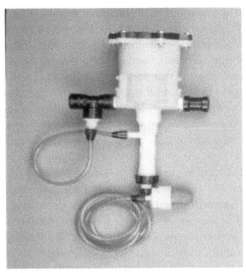

FIGURE 6.3   Hydraulic dosing pump or nonelectric proportional injector.

6.1.1   FERTIGATION INJECTION EQUIPMENTS

Chemicals for fertigation should cause no or minimum clogging of the emitters and drip irrigation systems. Water soluble fertilizers may be injected into the micro irrigation system by use of venturi suction devices, fertigation tanks operating on the principle of differential pressure, or injection pumps.

Venturi type injector delivers the fertilizers at a concentration, which depends on the water flow (Fig. 6.1). Thus the injection method is inaccurate because the pressure and flow rates vary in drip irrigation system. Venturi devices are popular and most ideal for small micro irrigation systems because of their simplicity, low cost, and no need of additional power source. Venturi devices generally have a significant head loss across them. The injection rate of these devices is low and is dependent on the system pressure and discharge. This will tend to limit their use on large systems and for crops with high fertigation requirements and high injection rates. This difficulty can be overcome by the installation of the series of venturi with a small centrifugal pump in a parallel circuit. The venturi injection method uses a venturi device to cause a reduced pressure (vacuum) that sucks the fertilizer solution into the line. A vacuum is created as the water flows through a converging passage that gradually widens. The device operates when there is a pressure differential between the water entering the injector and the fertilizer solution leaving into the irrigation system.

Fertilizer Tank containing fertilizer solution is connected to the irrigation pipe at the supply point. Part of the irrigation water is diverted through the tank diluting the nutrient solution and returning to the main supply pipe (Fig. 6.2). The concentration of fertilizer in the tank thus becomes gradually reduced. Fertilizer tank operates on the principle of differential pressure across them with the help of a throttling valve. These are simple steel tanks and are effective as well as reliable injection devices. They are used where higher dosages of chemical or fertilizer application are required without regard to the concentration of the chemical/fertilizer. These fertilizer tanks have their inlet and outlet connected to the main line at two points having different water pressures, across a simple throttling valve or a pressure-reducing valve. This causes the water to flow through the injector, gradually displacing the chemical it contains. Thus, the concentration of the applied chemical decreases continuously, and the chemical is gradually diluted until it has all been discharged into the irrigation system. The chief disadvantage is that the chemical concentration decreases with time. On small irrigation systems, the decreasing concentration may be unimportant and hence use of fertilizer tank for injection is ideal and simpler compared to use injection pumps. In fertilizer tank, the fertilizer is applied in one pulse after a certain application of water without fertilizer.

Hydraulic dosing pump or nonelectric proportional injector (Fig. 6.3) has an injection rate that is factory preset or may be chosen by an adjustable setting and injector body. This means they achieve an injection rate proportional to the flow of water passing through. The resulting solution strength is therefore constant, even if the water flow varies. In fertigation pump, the same dose of fertilizers is applied like in fertilizer tank method but in proportion to the water applied. The irrigation water has a fixed concentration of the applied fertilizer

Following formulae are useful in different situations:

1. Amount of fertilizer (W, kg) = [Qs × V × C]/[1000 × E × qi]   (1)

2. Concentration of fertilizer in the irrigation water (C, ppm) =
    [1000 × W × E × qi]
$$\frac{[1000 \times W \times E \times qi]}{[Q \times V]} \tag{2}$$

3. Required container tank capacity (V, liters) =
$$[1000 \times W \times E \times qi]/[Qs \times C] \tag{3}$$

4. Required injection rate (qi, lph) = [Qs × C × V]/[1000 × W × E]   (4)

where: W = amount of fertilizer (kg); Qs = sectional flow of the irrigation system (m /hr.); qi = Injection rate (lph); C = concentration of fertilizer in the irrigation water (ppm); V = Container volume (liters); and E = concentration of the nutrient in the fertilizer (fraction).

When pressurized irrigation systems are used, fertigation is not OPTIONAL but ABSOLUTELY NECESSARY. What happens if fertilizers are applied separately from the water? In drip irrigation, only ~30% of the soil is wetted by the drippers. Fertilization efficiency decreases because the nutrients will not be dissolved in the dry zones where the soil is not wetted. The benefits of irrigation will not be expressed. Therefore, fertigation is the only method to correctly apply fertilizers to irrigated crops.

Conventional preplant fertilizer: Plants get a larger dosage of fertilizer than they require at the time it is applied. Fertilizer losses occur. But in fertigation, fertilizers are applied in small doses initially and increased gradually matching the crop growth and demand according to nutrient uptake pattern directly to root zone of the crop resulting in higher fertilizer use efficiency compared to conventional soil application of fertilizers.

Fertigation program includes nutrient rates and nutrient ratio. The following recommendations are considered for nutrient rates:

- When a crop is first seeded/transplanted, begin with small amounts of fertilizer as the crop is growing slowly.
- Increase the amounts as the growth rate increases.
- As the crop matures and growth slows, reduce the fertilizer rates accordingly.
- For most crops, it is sufficient to program no more than four or five rate changes during the production season.

Nutrient ratio [8] takes into account following recommendations:

- Early in the crop's season, apply P and K in small doses for best rooting and plant establishment.
- When plants are vegetative, apply adequate quantities of N for best plant growth and development.
- As fruit load grow, increase K for best fruit setting, development and quality. High absorption of K is noticed during heavy fruit loads of vegetable crops and a massive stalk development of flower crops.
- Reduce N doses to avoid lush growth, soft fruit and pests' problems [8].

6.1.2 METHODS OF FERTIGATION

1. Proportional application
    - Delivers a constant ratio of nutrients applied to flow rate.
    - The injection rate is proportional to the water discharge rate, e.g. one liter of solution to 1000 L of irrigation water.
    - The fertilizer dose is expressed in concentration units (ppm).

2. Quantitative application
    - Application of the plant nutrients in predetermined concentrations to the irrigation system and the fertilizer concentration varies during its application.
    - Nutrient solution is applied in a calculated amount to each irrigation block, e.g. 20 L to block A, 40 L to block B, etc.
    - The fertilizer dose is expressed in kg per hectare.

6.1.3   INTERACTION BETWEEN FERTILIZERS AND OTHER FERTILIZERS (COMPATIBILITY)

The use of two or more fertilizer tanks allows to separate the fertilizers that interact and cause precipitation. Placing in one tank the calcium, magnesium and micronutrients, and in the other tank the phosphate and the sulfate compounds, enables safe and efficient fertigation.

6.1.4   REQUIREMENTS FOR FERTIGATION

- Test the native soil fertility status.
- Fix the correct fertilizer dose for selected crop.
- Develop appropriate fertigation schedule   .
- Select suitable fertilizer grade according crop stage.
- Install appropriate fertigation device along the main line.

6.1.5   STEPS IN FERTIGATION

- Calculate the required fertilizer quantity for the actual cropped area.
- Prepare the nutrient stock solution (dissolve the solid fertilizer with water at 1:5 ratio).
- Operate the drip system for 10–20 mts for wetting (1st process).
- Regulate the valves and initiate fertigation (2nd process) at 95 or 186 lph injection rate as per the fertigation device selected.
- Complete the fertigation and finally flushing (3rd process) for 10–15 mts for removal of left out fertilizers in pipe network.

6.1.6   DO'S AND DON'TS PRINCIPLES IN FERTIGATION

DO's	DON'ts
Use completely soluble fertilizers.	Fertilizers with impurities, low solubility and nutrients.
For P source use MAP or Ortho Phosphoric Acid.	Avoid Super Phosphate for fertigation.
Prefer White potash against Red Potash for K nutrient.	Avoid Chloride fertilizers for grapes and quality fruits, tobacco.
Select fertilizers with high nutrient contents.	Avoid fertilizers causing clogging, corrosion to fertigation devices and pipes.

DO's	DON'ts
Ensure correct drip fertigation design and schedule.	Improper design leads to wastage of fertilizer and water.
Use regulated emitters only.	Avoid tap and micro tubes.
Fertigation at more frequent intervals with low application rates. Fertigate during middle of irrigation to allow flushing of lines.	Avoid high rates of fertilizer per application and long intervals.

6.1.7 FERTILIZERS FOR OPEN FIELD CROPS AND FRUITS GROVES

For open field conditions where the yield and income is not equivalent to that of greenhouse conditions, integration of water soluble fertilizers (WSF) along with conventional urea and potash having higher solubility equivalent to WSF grades is an ideal practice to match fertilizer dose as well as to reduce the cost of fertilizers. The fertilizers requirements can be met from the following nutrient forms:

- N-NH NO and NH
- P-Phosphoric acid or MKP
- K-Potassium chloride

6.1.8 FERTILIZERS FOR GREENHOUSE CONDITIONS

Unlike open field conditions, the growing media is different, therefore, use of liquid fertilizers is warranted under greenhouse conditions. Investments are higher for greenhouses and use of liquid fertilizers are still costlier compared to regular WSF solid grades. Hence following nutrient forms (both for major and micro nutrients) are preferred under greenhouse conditions.

- N-NH NO
- P-Phosphoric acid or MKP
- K-Potassium Nitrate or MKP
- Micro-nutrients – Mg&B

6.1.9 CONCLUSIONS

Fertigation using high solubility NPK fertilizers produces higher yields with less loss of fertilizer than manual applications. Adoption of fertigation practice has increased rapidly worldwide since its introduction in the late 1960's. The uniform distribution and control of the water and nutrients makes fertigation an advanced and efficient fertilization practice. Nutrients are delivered to the restricted root zone (wetted soil) in a readily available form. Frequent delivery of water and nutrients replenishes the small volume of water and nutrients in the active root zone, nourishing the crop throughout the entire growing season. Fertigation can be achieved by using single or multiple nutrient fertilizers, in the solid or liquid form. An essential prerequisite for the use of solid fertilizer in fertigation is its complete and fast dissolution in the irrigation water. Fertigation allows nutrient needs to be supplied according to the physiological stage of the crop and the site-specific requirements (crop type and cultivar, growing system, yield target, soil type and climate conditions). Nutrient status should be monitored regularly in the plant tissue (lead or petiole analysis), in the soil and in the irrigation and leaching solutions.

6.2 RESEARCH ADVANCES IN FERTIGATION AT TAMIL NADU AGRICULTURAL UNIVERSITY (TNAU)

6.2.1. SUGARCANE WITH WATER SOLUBLE FERTILIZERS (WSF)

Figure 6.4 shows profit (Rs./ha) of drip irrigated banana, tomato and sugarcane with fertigation using water soluble fertilizers [32]. The results in Table 6.6 indicated that highest yield of sugarcane was recorded under drip fertigation with Water Soluble Fertilizer (WSF) at 75% NPK recommended dose (T3, 212.35 t/ha) when compared to control (surface irrigation + soil application of NF at 100% NPK dose (T5), 155.20 t/ha). This was followed by fertigation with WSF at 100% NPK recommended dose (206.65 t/ha) as indicated in Table 6.

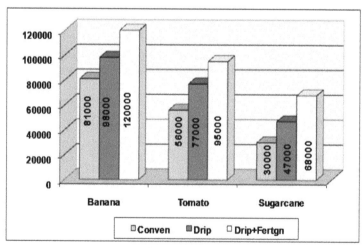

FIGURE 6.4   Profit (Rs./ha) of drip irrigated banana, tomato and sugarcane with fertigation using water soluble fertilizers [32].

TABLE 6.6 Fertigation in Drip Irrigated Sugarcane (Co. 853) With Water Soluble Fertilizers [4]

Treatments	Cane yield	Sugar yield	% water saving	Discounted B:C ratio	Net income
	t/ha	t/ha	%	ratio	Rs./ha
T1 – fertigation with WSF at 125% NPK recommended dose	208.90	31.42	28.95	6.39	55501
T2 – fertigation with WSF at 100% NPK recommended dose	206.65	30.69	28.95	6.69	58811
T3 – fertigation with WSF at 75% NPK recommended dose	212.35	31.80	28.95	7.48	67687
T4 – Drip irrigation + Soil appli-cation of NF at 100% NPK dose	172.15	25.25	28.95	5.90	49995
T5 – Surface irrigation + soil application of NF at 100% NPK dose	155.20	22.97	-	1.92	50415
SEd	0.983	-	-	-	-
CD (at P = 0.05)	2.143	-	-	-	-

B:C = Benefit–cost ratio.

The highest sugar yield of sugarcane was recorded under drip fertigation with WSF at 75% NPK recommended dose (T3, 31.80 t/ha) when compared to control (surface irrigation + soil application of NF at 100% NPK dose, 22.97 t/ha). This was followed by fertigation with WSF at 125% NPK recommended dose (31.42 t/ha).

6.2.2 SUGARCANE WITH CONVENTIONAL FERTILIZERS (NF)

The results of the study indicated that the application of 150% of recommended N and K in 14 equal splits registered a highest yield 183.9 Mt/ha but it was comparable with 125% of recommended N and K in 14 equal splits (Table 6.7). Fertigation at 100% dose has also registered a higher cane yield of 173.3 t/ha, while soil application at same dose gave only 135.3 t/ha (thus 28.09% increase over control). Similarly fertigation at 100% dose also recorded higher sugar yield (23.19 t/ha) with higher B:C ratio (1.82) as compared to control.

TABLE 6.7 Drip Fertigation With Conventional Fertilizers in Sugarcane [15]

Treatments	CCS%	Cane yield (t/ha)	Sugar yield (t/ha)	B:C ratio	% water saving
Drip fertigation with N&K at 100% dose in 14 splits	13.38	173.3	23.19	1.82	8.9
Control (surface irrigation and soil application of N&K at 100% dose)	12.84	135.3	17.37	1.78	-

6.2.3 BANANA WITH WATER SOLUBLE FERTILIZERS (WSF)

Drip fertigation in banana (var. Nendran) with WSF at 125% NPK (T1) registered the highest fruit yield of 42.65 t/ha, which was 66.76% increase over surface irrigation and soil application of normal fertilizers (T5, Table 6.8). Fertigation at 100% dose also registered higher fruit yield of 35.075 t/ha compared to control (T5, 25.58 t/ha).

TABLE 6.8 Drip Fertigation With Specialty Water Soluble Fertilizers in Banana (Nendran) [5]

Treatments	Yield (t/ha)	% water saving	B:C ratio	Net income Rs./ha
Drip fertigation with WSF at 125% NPK	42.650	35.23	1.96	134768
Drip fertigation with WSF at 100% NPK	37.450	35.23	1.98	117804
Drip fertigation with WSF at 75% NPK	35.075	35.23	2.27	114991
Drip irrigation + soil application of NF at 100% NPK	30.325	35.23	2.01	94741
Surface irrigation + soil application of NF at 100% NPK	25.575	-	1.71	80791

6.2.4 BANANA WITH CONVENTIONAL FERTILIZERS (NF)

Fertigation in banana (cv. Robusta) with conventional fertilizers like urea and potash with 25 L/day/plant (lpd) + 100:30:150 g NPK/plant has registered higher fruit yield of 95.00 t/ha in plant crop (Table 6.9). Thus there was an increased fruit yield of 61.07% with fertigation as compared to basin irrigation and soil application of conventional fertilizers (200:30:300 g NPK/plant).

TABLE 6.9   Fertigation With Conventional Fertilizers in Banana (cv. Robusta) [14]

Treatments	Bunch weight (kg)	Yield (t/ha)	% increase over conventional	B:C ratio
Plant crop 25 lpd + 100:30:150 g NPK/plant	38.00	95.00	61.07	1:1.78
Ratoon crop 25 lpd + 150:30:225 g NPK/plant	44.42	111.05	88.28	1:3.21
Control-Basin irrigation + 200:30:300 g NPK/plant soil application	23.59	58.98	-	1:0.93

6.2.5. HYBRID TOMATO WITH LIQUID AND WATER SOLUBLE FERTILIZERS (WSF)

Drip fertigation in hybrid tomato with WSF at 75% NPK applied drip through at 80% of 2 days CPE was found to be a superior technology, registering 26.0% increased fruit yields, 22.99% water saving, 25.0% saving in fertilizers, 27.57% increased fertilizer use efficiency (FUE) with higher B:C ratio of 3.47 over soil application of normal fertilizers (Tables 6.10–6.12).

TABLE 6.10   Comparison of Source of Fertilizers For Fertigation in Hybrid Tomato.

Treatments	Fruit yield (t/ha)	Water use (mm)	% water saving	% increase in FUE	B:C ratio
Liquid Fertilizer	62.394	390.6	22.99	25.67	2.48
Water soluble fertilizer	63.916	390.6	22.99	27.57	3.47
Normal fertilizer	50.603	390.6	22.99	-	3.06

TABLE 6.11   NPK Uptake By Tomato As Influenced By Fertigation With Method of Fertilizer Application [9]

Treatments	Nutrient uptake (kg/ha)		
	N	P	K
T1:control (soil application NF+Furrow irrigation)	109.3	9.5	69.1
T2: Soil application NF+ drip irrigation	142.1	13.3	94.3
T3: 100% fertigation with WSF	165.7	16.5	113.5
T4: 100% fertigation with NF	144.1	12.8	100.7
T5:75% RD WSF+ fertigation	140.5	12.9	92.3
T6: 1/2 soil +1/2fertigation	161.9	15.3	105.2
T7:NK fertigation + P soil	163.4	14.4	109.9
T8: sub surface drip fertigation	161.3	15.4	110.6
CD (p=0.05)	9.6	1.3	8.8

TABLE 6.12   Effects of WSF on Tomato Fruit Quality [9]

Treatments	TSS (Brix)	Titrable acidity (% citric acid)	Ascorbic acid (mg 100/fresh weight)
T1:control (soil application NF+Furrow irrigation)	3.95	0.4	16
T2: Soil application NF+drip irrigation	4.12	0.41	16.67
T3: 100% RD fertigation with WSF	4.28	0.46	19.33
T4: 100% RD fertigation with NF	4.15	0.43	17.33
T5:75% RD WSF+ fertigation	4.03	0.42	17.67
T6: 1/2 soil +1/2 fertigation	4.13	0.45	19
T7:NK fertigation + P soil	4.25	0.44	17.33
T8: sub surface drip fertigation	4.22	0.46	18
CD (p=0.05)	NS	0.04	1.17

6.2.6 FERTIGATION IN HYBRID COTTON WITH CONVENTIONAL FERTILIZERS (NF)

Fertigation through drip irrigation (100% N&K in 6 equal splits) in hybrid cotton (TCHB 213) increased the cotton yield (2367 kg/ha), which was 43.72% higher compared to surface irrigation and soil application of 100% NPK dose (Tables 6.13 and 6.14). Fertigation at 100% recommended dose of N and K as urea and potash applied in 4 equal splits (basal, 35 days after sowing, flowering and boll formation) and 6 splits (at 20 days interval from sowing) were found to be superior to conventional fertilization. Increasing the splits (6) gave B:C ratio of 1.82 compared to 1.72 for 4 splits.

TABLE 6.13   Fertigation With Conventional Fertilizers in Hybrid Cotton [17]

Treatments	Yield kg/ha	B:C ratio	% increase in yield
Fertigation with N&K at 100% dose in 4 equal splits	2239	2.49	35.94
Fertigation with N&K at 100% dose in 6 equal splits	2367	2.63	43.72
Fertigation with N&K at 75% dose in 4 equal splits	1892	2.20	14.87
Fertigation with N&K at 75% dose in 6 equal splits	2112	1.76	28.23
Control-Surface irrigation and soil application of NPK at 100% dose	1647	1.8	-
SEd	106	0.2	
CD	224	0.319	

TABLE 6.14   Effects of Drip Fertigation on Water Saving and Yield in Hybrid Cotton [33]

Methods	Water used mm	Water saving mm	Yield 100 kg/ ha	Additional yield 100 kg/ ha
Conventional	750	–	22	–
Drip fertigation	500	250 (33%)	30	8 (36%)

6.2.7   FERTIGATION THROUGH MICRO SPRINKLER WITH WATER SOLUBLE FERTILIZERS (WSF)

Microsprinkler fertigation of radish once in 2 days with 75% NPK dose with WSF like Mono Ammonium Phosphate (MAP) and Potassium Nitrate (multi-K) resulted 41.93 t/ha, which was 64.24% increase over control (surface irrigation and soil application of normal fertilizers at 100% dose (Table 6.15). Through micro sprinkler irrigation, there was water saving of 62.42% compared to surface irrigation (control). Micro sprinkler fertigation once in 2 days at 75% dose with WSF has also registered higher net income of Rs.76,582 with a high discounted B:C ratio of 9.47. Thus an additional net income of Rs.24,044 was realized compared to surface irrigation and conventional fertilization.

Performance of drip irrigated okra, chilies, egg plant, and coconut crops are compared in Tables 6.16–6.19 for fertigation and conventional methods at TNAU, Coimbatore.

TABLE 6.15   Fresh Root Yield of Radish Under Microsprinkler Fertigation [3, 23]

Treatments	Fresh radish yield t/ha	% increased yield over control	Net income Rs/ha	Discounted B:C ratio
T1- MSF 100% WSF – 2 days	42.31	65.73	75868	9.39
T2 – MSF 75% WSF – 2 days	41.93	64.24	76582	9.47
T3 – MSF 50% WSF – 2 days	37.52	46.96	67198	8.48
T4 – MSF 100% WSF – 4 days	39.53	54.83	68918	8.66
T5 – MSF 75% WSF – 4 days	37.24	45.86	64857	8.24
T6 – MSF 50% WSF – 4 days	34.58	35.45	59848	7.79
T7 – MSI 100% NF – 2 days	35.51	39.09	63917	8.18
T8 – MSI 100% NF – 4 days	32.10	25.73	55394	7.24
T9 – Control	25.53	–	52538	4.65
SEd	0.76	–	–	–
CD (P=0.05)	1.60	–	–	–

TABLE 6.16 Yield Attributes and Fruit Yield of Okra As Influenced by Drip Fertigation With WSF [3]

Treatments	Number of fruits per plant	Weight of single fruit g	Fruit yield t/ha
T1-Drip 100% + fertigation100% NPK-WSF	17.7	15.11	19.83
T2 Drip 100% + fertigation 80% NPK-WSF	17.1	15.05	19.10
T3Drip 100% + fertigation 60% NPK-WSF	17.6	14.25	18.56
T4 Drip 75% + fertigation100% NPK-WSF	18.5	15.32	20.97
T5 Drip 75% + fertigation 80% NPK-WSF	18.2	15.97	21.50
T6 Drip 75% + fertigation 60% NPK-WSF	17.8	14.60	19.25
T7 Drip 50% + fertigation100% NPK-WSF	15.7	13.75	15.97
T8 Drip 50% + fertigation 80% NPK-WSF	15.1	13.25	14.84
T9 Drip 50% + fertigation 60% NPK-WSF	15.6	12.22	14.14
T10 Furrow 1.0 IW/CPE + soil application 100% NPK-NF (soil)	15.0	13.45	14.94
SEd	0.3	0.75	0.34
CD (P=0.05)	0.6	1.58	0.72

TABLE 6.17 Effects of Drip Fertigation on Water Saving and Yield in Chilies [33]

Method	Water used, mm	Water saving, mm	Yield of green chilies, t/ha	Additional yield, t/ha
Conventional	700	-	22	-
Drip fertigation	450	250 (35%)	35	13 (59%)

TABLE 6.18 Fertigation of Drip Irrigated Egg Plant [33]

Method	Water used, mm	Water saving, mm	Yield t/ha	Additional yield, t/ha
Conventional	750	-	40	-
Drip fertigation	550	200 (26%)	75	35 (87%)

TABLE 6.19 Effects of Drip Fertigation on Water Saving and Yield of Coconut [30–33]

Methods	Water used, lpd/tree	Water saving, lpd/tree	Yield nuts/ tree/year	Additional yield, nuts/tree/year
Conventional Method	185	–	108	–
Drip fertigation	100	85 (45%)	156	48 (44%)

6.2.8. ECONOMICS OF MICRO IRRIGATION AND FERTIGATION

The cost for installing drip irrigation varies from Rs.30,000 to 45,000/ha for wide spaced crops like coconut, mango etc. to Rs.50,000 to 95,000/ha for closely spaced crops like sugarcane, cotton, vegetables, etc. The cost of the system depends upon the crop type, row spacing, crop water requirement, and distance from water source, etc. The economics of micro irrigation has been calculated with and without fertigation (Table 6.20). It can be observed that the payback period is about one year for most of the crops and the benefit cost ratio varies from 2 to 5.

Under precision farming project at Dharampuri District in Tamil Nadu (TNPFP), the researchers evaluated economic impacts of four crops (tomato, chilies, cabbage and cauliflower). The results are summarized in Table 6.20 for a 100-hectare farm. The gross income realized varied from Rs.3.90?10 Rs./ha for tomato compared to Rs.1.72?10 for chili. Of the four crops, tomato gave a maximum gross income of Rs.3.90?10. The entire gross income was considered a net income for farmers in the production, since 100% subsidy was provided by the government.

TABLE 6.20   Crop Yields in TNPFP and Nonproject Area [28]

Crop	TNPFP, 100 ha	nonproject area	
	Yield t/ha		% increase in yield
Cabbage	60	50	20
Cauliflower	60	50	20
Chilies	29	15	95
Tomato	65	40	63

6.2.9   CONCLUSIONS

Surface irrigation will still dominate as the primary irrigation method in India. However, with the current trends, the area under micro irrigation will continue to expand [12]. Precise and efficient use of water and nutrients are the prime concerns of sustainable crop production in India. Fertigation is a sophisticated and efficient method of applying fertilizers, in which the irrigation system is used as a carrier and distributor of the crop nutrients. Correct design of micro irrigation is an essential prerequisite for efficient distribution of nutrients thus avoiding deficiency in some pockets and excess in other areas. Various types of fertilizers meant for open field and greenhouse conditions can be chosen for appropriate use for achieving maximum fertilizer use efficiency. Fertigation schedule should be developed before stating fertigation, considering essential criteria like: native soil fertility status, targeted yield, variety or hybrid, growing conditions, nutrient uptake pattern, actual soil and plant nutrient concentrations etc. The fertigation with water-soluble fertilizers is a costly process owing to high cost of water-soluble fertilizers. Hence fertigation with imported grades of WSF have to be targeted in high value crops for getting greater net return and to have shorter payback periods. The synergism and combination of water and nutrient lead to an efficient use of both by the plant. Based on the studies conducted on different horticultural crops, it was found that adoption of this technology improves the yield and the quality of crops. It is also highly beneficial to farming community to reduce the cost of production. Further, it helps in maintaining the soil health for better productivity and reducing environmental pollution.

6.3 FERTIGATION WITH WATER SOLUBLE FERTILIZERS IN BANANA (CV. NENDRAN)

6.3.1 INTRODUCTION

Banana ( ) is the fourth most important global food commodity after rice, wheat and milk in terms of gross value of production. Banana is a globally important fruit crop with 97.5 million tons of production. With total annual production of 16.91 million tons from 490.70 thousand ha with national average of 33.5 t/ha. Maharashtra ranks second in area and first in production with 60 t/ha. Banana contributes 37% to total fruit production in Indi. Bananas occupy 20% area among the total area under crop in India. Banana requires plentiful supply of water for higher production. The total ET demand of banana is about 1600 mm. This is of particular importance since the available amount of water in the world is only about 1520 million cubic kilometers: 97% is ocean water and sea salt, 2% is frozen arctic waters and only 1% is water lakes, rivers and underground water, which is potable water for direct use to humans [22].

The efficient use of water is important under limited water resource conditions. Drip irrigation (trickle or micro irrigation) is a promising system for economizing the use of available irrigation water. It is also necessary to manage the available water efficiently for maximum crop production. Drip irrigation can apply water both precisely and uniformly at a high irrigation frequency compared with furrow and sprinkler irrigation, thus potentially increasing yield, reducing subsurface drainage, providing better salinity control and better disease management, since only the soil is wetted whereas the leaf surface stays dry. The goal of micro irrigation is to minimize water use and to increase the efficiency of applied fertilizers. Adoption of drip irrigation might help in increasing the irrigated area, productivity of crops and water use efficiency. The drip irrigation once in 2 days at 24 L/plant has resulted in 41.6% water saving in Tamil Nadu conditions [1]. In Karnataka, drip irrigation at 80% of cumulative pan evaporation (CPE) has brought 25% saving in irrigation water. The yield increase in drip irrigation compared to conventional irrigation method varies from 20 to 100%, whereas saving in water ranges from 40 to 70%besides 50 to 60% saving in labor cost [24].

Fertigation is yet another superior technology to apply of fertilizers through drip irrigation system for achieving higher water and fertilizer use efficiencies. Application of fertilizers through drip irrigation system (fertigation) can reduce fertilizer usage, minimize leaching by rain and excessive irrigation, maximize the fertilizer use efficiency, allows flexibility in timing fertilizer application, and reduces the labor cost for applying fertilizer. Increased growth and yield with drip fertigation has been reported in several crops and the yield increase ranged between 7–112% depending on the crops, varieties and methods of irrigation. The water and fertilizer saving through drip fertigation system have been reported to be 40–70 and 30–50%, respectively [20].

In this section, the research studies are discussed: To evaluate the effects of drip fertigation on yield of banana; To study water use efficiency and nutrient use efficiency of Banana under drip fertigation compared to conventional practices by the farmer.

6.3.2 MATERIAL AND METHODS

The experimental location was in Puttuvikki of Sundakkamuthur village of Coimbatore located 7 km away from TNAU campus on the southwestern side. The village has a tank, which gets filled up partially during Northeast monsoon period of September through December. The latitude is 11°N and 77°E longitude at an altitude of 427 m above MSL. The average annual rainfall of Coimbatore is 640 mm, received in 47 rainy days. The mean annual maximum and minimum temperatures were 31.4 °C and 21.2 °C. The soil at the experimental field is well-drained clay loam medium in available N, high in available P and K. The pH of the soil is moderately alkaline but the EC is normal. The irrigation water has mild alkaline pH and the EC is normal. The other cations like Ca, Mg, and anions like anions like CO and HCO are in the normal range.

Treatments consisted of:

T1 – Drip fertigation with WSF at 125% dose once in 2 days;
T2 – Drip fertigation with WSF at 100% dose once in 2 days;
T3 – Drip fertigation with WSF at 75% dose once in 2 days;
T4 – Drip irrigation + Soil application of NF at 100% dose once in 2 days;
T5 – Conventional irrigation at 1.0 IW/CPE + soil application of NF (Normal fertilizer: Control) at 100% dose.

Banana suckers were planted at 2×2 m spacing. The 12 mm laterals were laid out at 2 m apart and drippers of 8 lph were at 2 m interval. The laterals were taken off either side from the 40 mm PVC submain running at center of the main field. Here, the submain was connected to 63 mm PVC mainline. Each plot was provided with 40 mm flushing valve. One venture injector was fitted at the field head succeeded with a 2″ mesh filter. For banana, fertigation schedules were prepared with different fertilizer grades according to growth stages and crop requirements. Computed fertilizers quantities were dissolved at one part of fertilizer with five parts of irrigation water and nutrient stock solution was prepared. At every fertigation, drip system was run for wetting as a first step and then fertigation was done and finally flushing was done 5–10 min once in 2 days. Note: Always fertigate during the middle third of irrigation duration.

6.3.2.1   IRRIGATION SCHEDULING

Drip irrigation was scheduled based on evaporation rate once in 2 days from the USWB Class A open pan evaporimeter. The irrigation water requirement through drip was computed according to equation (5).

$$V = [2 \text{ days CPE}] \times Kp \times Kc \times A \times Wp \qquad (5)$$

where: V = Volume of irrigation water in liters/plant/once in 2 days; CPE = cumulative pan evaporation once in two days; Kp = pan coefficient (= 0.75); Kc = crop coefficient; A = area = $2 \times 2 = 4$ m; and Wp = wetted percentage = 40%. The depth of irrigation thus varied according to evaporation rate and crop factor at initial, crop development, mid-season and late season stages of the crop. Irrigation time was calculated based on volume of water to be irrigated in relation to number of emitters per plant and the discharge rate of emitters.

6.3.3   RESULTS AND DISCUSSION

6.3.3.1   YIELD PARAMETERS

6.3.3.1.1   NUMBER OF FINGERS/BUNCH

Drip fertigation of banana had significant influence on the number of fingers per bunch. Among the treatments, significantly higher number of fingers per bunch was recorded under drip fertigation with water soluble fertilizers at 125% NPK dose (= 61.20) and this was on par in drip fertigation with water soluble fertilizers at 100% NPK dose (= 59.41) compared to significantly lower number of fingers/bunch under surface irrigation + soil application of normal fertilizers (= 47.65). Similarly in second crop, significantly higher number of finger/bunch was observed under drip fertigation with water soluble fertilizers at 125% NPK dose (= 59.85), which was at par with drip fertigation with water soluble fertilizers at 100% NPK dose (= 58.67).

6.3.3.1.2   NUMBER OF HANDS/BUNCH

The number of hands per bunch was significantly influenced by drip fertigation in banana. Drip fertigation with water soluble fertilizers at 125% NPK dose recorded significantly higher number of hands/bunch (= 5.50 and 5.42) followed by drip fertigation with water soluble fertilizers at 100% NPK dose (= 5.30 and 5.05) in both crops, respectively. Surface irrigation and soil application were found to be inferior with only 4.63 hands/bunch. The increase in number of hands /bunch were 18.70% and 14.40% under drip fertigation at 125% and 100% RDF with water soluble fertilizers, respectively compared to surface irrigation with soil application of normal fertilizers.

6.3.3.1.3 FINGER WEIGHT

The maximum finger weight (= 0.275 and 0.262 kg) were recorded under drip fertigation at 125% RDF with water soluble fertilizers in both the crops, respectively. The minimum finger weight (= 0.215 and 0.208 kg) were recorded in surface irrigation with soil application of normal fertilizers, as shown in Table 6.21.

TABLE 6.21 Effects of Drip Fertigation of Banana on Number of Fingers/Bunch, Number of Hands/Bunch and Finger Weight

Treatment	Number of fingers/bunch		Number of hands/bunch		Finger weight (kg)	
	First crop	Second crop	First crop	Second crop	First crop	Second crop
T1	61.20	59.85	5.50	5.42	0.275	0.262
T2	59.41	58.67	5.30	5.05	0.270	0.255
T3	54.30	53.91	4.83	4.65	0.256	0.245
T4	53.45	53.05	4.15	4.28	0.250	0.232
T5	47.65	46.28	4.63	4.54	0.215	0.208
SEd	1.181	1.24	0.187	0.19	0.003	0.004
CD (at P = 0.05)	2.573	2.49	0.407	0.39	0.007	0.009

The drip irrigation increased bunch weight, fruit diameter and length, number of hands per bunch and number of leaves at flowering and at harvest compared to surface irrigation due to better utilization of water and nutrient by application of water and nutrients in small quantities directly in the root zone of crop based on crop requirement [26]. In surface irrigation, all yield-attributing characters were found minimum compared to drip method [10, 11].

6.3.3.1.4 BUNCH WEIGHT

Significantly higher bunch weight was recorded under drip fertigation with water soluble fertilizers at 125% NPK dose (= 17.06 kg and 16.75 kg) and this was at par with drip fertigation with water soluble fertilizers at 100% NPK dose (= 16.42 kg and 15.81 kg) in I and II crops, respectively. Next to this drip fertigation with water soluble fertilizers at 75%, NPK dose was also superior in registering higher bunch weight. Whereas, lesser bunch weight of 10.23 and 10.05 kg/bunch was recorded in surface irrigation with soil application of normal fertilizers in both the crops, respectively. The increase in bunch weight was 66.66 and 57.31%compared to surface irrigation with soil application of normal fertilizers in both the crops, respectively (Fig. 6.5). Increased bunch weight under drip irrigation compared to basin irrigation was also associated with the corresponding significant increase in the number of hands, total number of fingers, finger weight, length and circumference and pulp-peel ratio [11].

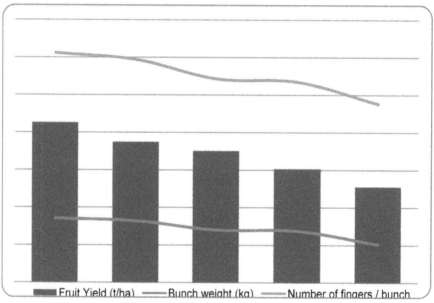

FIGURE 6.5   Relationship between fruit yield and yield parameters under drip fertigation in banana (Solid bars = Fruit yield in t/ha; Top curve = No. of fingers per bunch; and lower curve = Bunch weight).

6.3.3.1.5   FRUIT YIELD

Drip fertigation with water soluble fertilizers exerted significant and favorable influences on fruit yield of Nendran Banana. Significantly higher fruit yield of 42.65 t/ha and 41.42 t/ha were registered in drip fertigation with water soluble fertilizers at 125% NPK dose in the both crops, respectively; and next to this treatment drip fertigation with water soluble fertilizers at 100 and 75% doses registered higher fruit yields in the both crops. Conventional practice of surface irrigation + soil application of 100% normal fertilizers recorded the lowest yield of 25.58 t/ha and 24.12 t/ha in both the crops, respectively. Thus drip fertigation was superior in accounting for 66.73 (125% RDF) and 46.40 (100% RDF) percent increase in fruit yield over control in first crop. While, in second crop, similar trend was seen (Table 6.22). As there is tremendous response to fertigation in banana especially for N& K nutrients, higher dose of 125% NPK level registered the top most yield under fertigation. The increased banana yield of 41,000 kg/ ha was registered under drip fertigation which was 31% superior compared to surface irrigation [27]. The surface drip irrigation increased fruit yield of both main and ratoon crop of banana by 9.12 and 12.85%, respectively [30–33]. It was also observed that the drip method is superior to surface method of irrigation in terms of yield, quality, water saving and cost economics [19].

TABLE 6.22   Finger Weight and Bunch Weight in Nendran Banana as Influenced By Drip Fertigation

Treatments	Fruit Yield (t/ha)		Fruit Yield (t/ha)	
	First crop	Yield increase over control (%)	Second crop	Yield increase over control (%)
T1	42.65	66.73	41.42	71.72
T2	37.45	46.40	35.98	49.17
T3	35.08	37.14	34.15	41.58
T4	30.33	18.57	29.56	22.55
T5	25.58	-	24.12	-
SEd	0.63	-	0.58	-
CD (0.05)	1.26	-	1.17	-

6.3.3.2   WATER REQUIREMENTS FOR DRIP IRRIGATED BANANA

In first crop of banana, total water under drip irrigation once in 2 days was 1153.4 mm (drip irrigation requirement + effective rainfall). Under surface irrigation, 2200 mm of water was used. In second crop of banana, the total water used was 1177.4 mm compared to 1950 mm under surface irrigation. Thus water saving was 47.57 and 39.63% under drip fertigation compared to surface irrigation in first and second crops, respectively (Table 6.23). The water saving to the extent of 20% [18] and 50% [19] under drip irrigation in banana was achieved compared to surface method. Drip irrigation at 80% of evaporation replenishment was able to bring about nearly 25% saving in irrigation water in banana [25].

TABLE 6.23   Effect of Drip Fertigation on Water Saving in Banana

Treatments	First crop		Second crop	
	Water used (mm)	Water saving (%)	Water used (mm)	Water saving (%)
T1	1153.4	47.57	1177.4	39.62
T2	1153.4	47.57	1177.4	39.62
T3	1153.4	47.57	1177.4	39.62
T4	1153.4	47.57	1177.4	39.62
T5	2200	-	1950.0	-

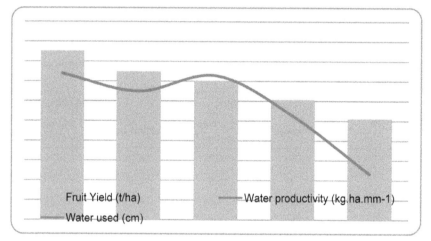

FIGURE 6.6   Variation of water requirement, fruit yield and water use efficiency under drip fertigation in banana among five treatments (Top curve: Water use efficiency; Solid bars = Fruit yield).

6.3.3.2.1   WATER PRODUCTIVITY (WUE)

Drip fertigation is an efficient method to deliver water and nutrients to the root zone of plants, because water is directly applied in the effective root zone of crop. Drip fertigation levels indicated profound influence on water use efficiency in banana (Fig. 6.6). The maximum water productivity of 36.97 and 35.18 kg/(ha.mm) were registered under drip fertigation with water soluble fertilizers at 125% NPK dose in both the crops, respectively, and followed by drip fertigation with water soluble fertilizers at 100% NPK dose. This indicated that there is an increased response of banana for higher drip irrigation levels. Water productivity was almost doubled under drip fertigation at 75% dose compared to soil application of fertilizers. The increased water use efficiency recorded under drip fertigation system was mainly due to better performance of the crop and increased yield by effective utilization of available water and nutrients that were supplied at regular intervals throughout the crop period to meet the crop demand [7].

6.3.3.3   FERTILIZER PRODUCTIVITY OF DRIP IRRIGATED BANANA

In contrast to yield levels, the highest fertilizer productivity (34.646 and 33.728 kg. fruit per kg NPK.ha) was associated with drip fertigation with water soluble fertilizers at 75% NPK dose in both crops, respectively and followed by drip fertigation with water soluble fertilizers at 100% NPK dose with 27.740 kg fruit per kg NPK. ha in 1st crop of banana (Fig. 6.7). Plants were taller (3%) and flowered 15 days earlier under the drip irrigation than the basin irrigation. Fertigation saved from 20% to 50% in fertilizers, while improving the yield and quality compared with the common methods of fertilizer application [16]. Application of urea through irrigation system was more efficient than hand broadcasting on soil surface on banana. More yield and significantly higher number of hands per bunch were obtained with fertigation [2]. Fertigation saved upto 25% of soluble fertilizers like urea, sulfate of potash through drip irrigation [25].

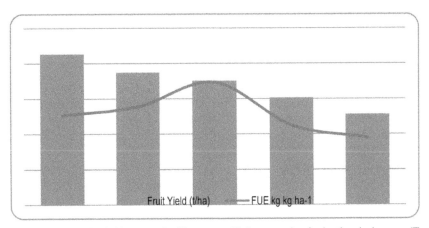

FIGURE 6.7   Fruit yield versus fertilizer use efficiency under fertigation in banana (Top curve: Fertilizer use efficiency; and solid bars = Fruit yield in t/ha).

6.3.3.4   ECONOMIC FEASIBILITY OF DRIP IRRIGATED BANANA

The net income (Rs./ha) was the highest in drip fertigation with water soluble fertilizers at 125% dose (Rs.241,393/ha) followed by Rs.211,429/ha and Rs.202,691 under 100 and 75% dose, respectively. Obviously lower net seasonal income was obtained under conventional method (Rs. 144,741/ha) due to lower yield levels. The economic analysis was proceeded further considering the water saving benefits under drip system. Due to drip irrigation in banana, 47.57% water saving resulted an additional area coverage of 0.48 ha under drip irrigation. Thus, computed gross income from overall area of 1.48 ha was Rs. Rs.472,048/ha under 125% dose compared to Rs.388,264 under 75% dose fertigation level. The marginal benefit: cost analysis showed that due to high cost of water soluble fertilizers drip fertigation at lower dose of 75% registered higher extra income per extra rupee invested (4.67 and 4.75) in I and II crops, respectively, while it was 4.08 and 4.05, respectively, under drip fertigation at 125% dose.

6.3.3.5 CONCLUSIONS

Banana is a major cash crop in India, yield and crop quality being critically dependent on supply of water and nutrients. Based on experiments during two season, it was concluded that drip fertigation with water soluble at 125% were possible to achieve more than two-fold higher water use efficiency, and at the same time reduced fertilizer requirement and raised crop productivity. Bunch yields of banana increased by 35 to 72% under drip fertigation with water soluble fertilizers. This was achieved in addition to 25 to 48% water saving with higher fertilizer and water use efficiency. Considering the economics affordable, farmers were able to adopt drip fertigation at 125%dose in Nendran Banana for overall higher fruit yield and gross returns, while for farmers with resource constraints, 75% dose was more economical.

6.4 SUMMARY

This chapter briefly discusses the fertigation technology in micro irrigation under Indian conditions. Research advances in fertigation of drip irrigated crops are also presented. The chapter also presents the effects of drip fertigation regime with water soluble fertilizer (WSF) for maximizing the fruit yield, water and fertilizer use efficiency of Banana (Nendran).

Water soluble fertilizers were tried at three doses viz., 125, 100 and 75% of recommended NPK by drip fertigation. These were compared with drip irrigation and soil application of Normal fertilizers (NF) as well as farmers' practice of surface irrigation and soil application of NF at 100% recommended dose. Banana suckers were planted at 2 × 2 m spacing).

Imported water soluble fertilizers viz., Mono Ammonium Phosphate (12:61:00), Polyfeed (19:19:19), Potassium Nitrate (13:00:45) and Urea were tried under fertigation. Drip irrigation was scheduled based on cumulative pan evaporation rates once in 2 days considering the pan and crop coefficient values at various stages. The crop factor (Kc) values considered were 0.4, 1.2 and 0.8 at planting, bunch development and maturity stages, respectively. Fertigation was scheduled once in 2 days with varying forms of fertilizer grades according to nutrient requirement at varying growth stages.

Results from the investigation showed that banana responded well to different fertigation regimes on yield and quality. The total water used under drip irrigation once in 2 days was 1153.4 mm while under surface irrigation it was 2200 mm, thus drip irrigation in Nendran banana has resulted in water saving of 47.57%. The water requirement of banana was precisely computed and applied as point source of irrigation directly in the rhizosphere, leading to higher yield per unit of water applied. This has resulted higher water use efficiency of 36.97 kg/(ha.mm) compared to surface irrigation (farmers' method, 11.62 kg/(ha.mm)).

Among different fertigation regimes, the yield contributing characters of banana were favorably improved at higher fertigation regime (125% dose with WSF) compared to lower regimes of 100 and 75%. Higher number of hands (5.5 per bunch), number of fingers (61.20), finger length (22.58 cm), finger girth (14.21 cm), finger volume (254.40 cm), pulp/peel ratio (4.28) and finger weight (0.275 kg) were registered under higher fertigation regime of 125% NPK with WSF compared to soil application of NF at 100% dose. In terms of banana productivity, maximum fruit yield of 42.65 t/ha was achieved by drip fertigation with 125% dose which was 66.73 and 40.62% increase over farmers' method (25.58 t/ha) and drip irrigation with soil application of NF at 100% dose (30.33 t/ha), respectively. Higher fertigation regime also gave higher net seasonal income of Rs.241,393 (US $4494/ha) over farmer's method (Rs.144,714 = US $2695/ha). It was concluded that drip fertigation with WSF at 125% dose of fertilizer in Nendran banana resulted higher water saving, fruit yield water and fertilizer use efficiency.

KEYWORDS

- **Banana**
- **CEP**
- **check basin**
- **chemigation**
- **cotton**
- **drip irrigation**
- **eggplant**
- **emitter**
- **fertigation technology**
- **fertilizer use efficiency, FUE**
- **ICID**
- **maize**
- **okra**
- **point source**

- radish
- rhizosphere
- sugarcane
- surface irrigation
- TNAU
- tomato
- venturi
- venturi injector
- water productivity
- water saving
- water soluble fertilizers, WSF
- Water Technology Center India, WTC
- water use efficiency, WUE

REFERENCES

1. Anonymous, (2003). Annual report of All India Coordinated Research Project on Water Management, WTCER, Bhubaneshwar, Orissa. Pages 64.
2. Arscott, T. G. (1970). Nitrogen fertilization of banana (Musa cavendishhii Lambert) through a sprinkler irrigation system. Trop. Agric. Trin., 7(1), 17–22.
3. Asokaraja, N., Muthukrishnan, P. (2012). Evaluation of water soluble fertilizers on the yield of Bhendi under drip fertigation system. In Proc. of the All India Seminar on Engineering Interventions for Profitable Agriculture held at AEC&RI, Sep 11–12.
4. Asokaraja, N. (2002a). Maximizing the productivity and quality of banana and sugarcane with water soluble fertilizers through drip fertigation. Annual Report. 2001–2002 WTC, TNAU, Coimbatore
5. Asokaraja, N. (2002c). Performance evaluation of drip fertigation system with liquid and water soluble fertilizers for increasing the yield and quality of vegetable crop (Tomato), Final Report of ICAR Adhoc Project 1999–2002, WTC, TNAU, Coimbatore.
6. Asokaraja, N., Selvaraj, P. K., Mahendran, S., Palanisamy, K. (2005). Fertigation Technology. Technical Bulletin, Tamil Nadu Agricultural University, Coimbatore
7. Bangar, A. R., Chudhari, C. (2004). Nutrient mobility in soil, uptake, quality and yield of suru sugarcane as influenced by the drip fertigation in medium vertisols. J. Indian. Soc. Sci., 52:164–171.
8. Hagin, J., Sneh, M., Lowengart–Aycicegi, A. (2003). Fertigation – Fertilization through irrigation. IPI Research topics No. 23, Switzerland.
9. Hebbar, S. S; Ramachandrappa, B. K; Nanjappa, H. V., Prabhakar, M. (2004). Studies on NPK drip fertigation in field grown tomato (Lycopersiconesculentum Mill.). , 21:117–127.

10. Hedge, D. M., Srinivas, K. (1990). Growth, productivity and water use of banana under drip and basin irrigation in relation to evaporation replenishment. Indian J. Agron; 35 (1&2), 100–102.
11. Hegde, D. M., Srinivas, K. (1991). Growth, yield nutrient uptake and water use of banana crops under drip and basin irrigation with N and K fertilization. Trop. Agric; 86(4), 331–334.
12. ICID, 2010. Annual Report 2009–2010. www.icid.org
13. IPI – FAI, 2003. Manual for First Training on Fertigation. Pune, Maharashtra Sep 15–16.
14. Kumar, N. (2003). Fertigation with conventional fertilizers in banana (cv. Robusta). Annual Report of Horticultural College and Research Institute, TNAU, Coimbatore.
15. Mahendran, S. (2003). Drip fertigation for maximizing the yield of sugarcane crop. Annual Report of ICAR Adhoc Project 2002–2003, ARS (TNAU), Ramanathapuram,
16. Malakouti, M. J. (2004). The Iranian experiences in fertigation and use of Potash fertilizers. IPI regional workshop on Potassium and Fertigation development in West Asia and North Africa. Rabat, Morocco.
17. Meyyazhagan, N. (2003). Effect of fertigation on the productivity of cotton (TCHB 213). All India Co-ordinated Cotton Improvement Project, Annual Report, 2002–2003, TNAU, Coimbatore.
18. More, M. R; Jadhav, S. N; Shinde, V. D., Jadhav, S. B., Pendke, M. S. (1999). Effect of varying depths of irrigation through drip under differential planting pattern on yield of banana. In Proc: All India Seminar on Micro irrigation –Prospects and Potential in India, pages 46, July 18, AP, Hyderabad.
19. Pawar, D. D; Rasskar, B. S., Bangar, A. R; Bhoi, P. G and Shinde S. H. (2000). Effects of water soluble fertilizers through drip and planting techniques on growth, yield and quality of banana. In Proc: International Conference on Micro and Sprinkler irrigation systems, 8–10 Feb, pages 515 –519, Jain Irrigation Hills, Jalgaon, Maharashtra.
20. Rekha, K; Mahavishnan, K. B. (2008). Drip fertigation in vegetable crops with emphasis on lady finger (Abelmoschusesculentus l. Moench). Agricultural Reviews, 29(3).
21. Satisha, G. C. (1997). Fertigation – New concept in Indian Agriculture. Kissan World, page 30. June.
22. Shaker, B. A. (2004). Effect of drip irrigation system on two varieties of Phaseolus bean production under the open field condition of Sudan. M.Sc. (Agric). Thesis, Omdurman Islamic University, Khartoum, Sudan.
23. Shobana, A., Asokaraja, N. (2002). Performance evaluation of micro sprinkler fertigation with water soluble fertilizers on water, fertilizer use and yield of radish. M.Sc.(Ag.) Thesis, Dept of Agronomy, TNAU, Coimbatore
24. Sivanappan, R. K. (2004). Irrigation and rainwater management for improving water use efficiency and production in cotton crop. In proceeding of International Symposium on "Strategies for sustainable cotton production. A global vision" 23–25 November, University of Agric. Sciences, Dharwad, Karnataka.
25. Srinivas, K. (1998). Standardization of micro irrigation in fruit crops- An overview. In Proc: Workshop on Micro irrigation and sprinkler irrigation systems 28–30 April, II-30, Central Board of Irrigation and Power, New Delhi, India.
26. Srinivas, K; Reddy, B. M. C., Chandra, S. S., Gowda, T., Raghupati, H. B., Padma, P. (2001). Growth, yield and nutrient uptake of Robusta banana in relation to N and K fertigation. Indian journal of Horticulture, 58(4), 287–293
27. Thadchayini, T., Thiruchelvam, S. (2005). An economic evaluation of a drip irrigation project for banana cultivation in Jaffna district. Water professional day symposium. Water Resources Research in Sri Lanka.

28. Vadivel, E; Ranghaswami, M. V., Manian, R. (2006). Tamil Nadu Precision Farming Project. Paper presented in the 7th International Micro Irrigation Congress, 13–15 September, Kuala Lumpur.
29. Walker, W. R. (1989). Guidelines for designing and evaluating surface irrigation systems. Irrigation and Drainage. Paper 45. FAO, Rome, Italy.
30. WTC, (1985–2003). Annual Reports of Water Technology Center, Tamil Nadu Agricultural University, Coimbatore.
31. WTC, (2003). Annual Report of Water Technology Center, Tamil Nadu Agricultural University, Coimbatore.
32. WTC, (2004). Annual Report of Water Technology Center, Tamil Nadu Agricultural University, Coimbatore.
33. WTC, (2008). Annual Report of Water Technology Center, Tamil Nadu Agricultural University, Coimbatore.

APPENDIX I

Photos of Experimental Site

CHAPTER 7

DESIGN OF AN EMITTER

MOHAMED EL-SAYED EL-HAGAREY

CONTENTS

Modified from Mohamed El-Sayed El-Hagarey, "Design and Manufacture of Pottery Dripper for the Use of Saline Water in Irrigation System," *IOSR Journal of Agriculture and Veterinary Science* (IOSR-JAVS), 7(5, May 2014), 70–80. http://iosrjournals.org/iosr-javs/papers/vol7-issue5/Version-4/M07547080.pdf. Modified by the author. Originally published via open access.

7.1   INTRODUCTION

Pottery dripper was designed and manufactured to allow the use of saline water in irrigation systems. In Egypt, 2 to 3 thousand million m^3 of saline water are for irrigating about 405,000 ha of land. The main objectives of this research study are to: (i) Design and develop a new pottery emitter to reduce saline water concentration; and (ii) manufacture and evaluate a pottery dripper.

Bhatt et al. [2] indicated that pitchers technique can be successfully employed at places with salinity and alkalinity problems. Salt deposition on the wall of pitchers does not adversely affect the plant growth as the plants continue to draw water from the pitchers. There is a high degree of correlation between the rate of water diffusion through small pitchers and large pitchers. Use of small-sized pitchers is more beneficial both in terms of water saving and in economic terms compared to large pitchers. Vasudaven et al. [6] indicated that daily depletion (%) from buried pitchers was slightly decreased with time. Mean daily depletion (%) was also decreased with the increase in salinity of the salt water. Salt distribution in soil around the pitcher was increased with the increase in horizontal distance from pitcher and was decreased on moving vertically downwards. Salinity of residual water in pitchers was increased with time showing that these pitchers have the capability to retain water. Mathai and Simon [4] concluded that water diffusion is more consistent with respect to the organic matter-mixed pottery samples rather than the sand-mixed ones.

Uniform porosities of organic matter-mixed pottery discs, as evidenced by optical micrographs, further exemplify that pots/pottery sheets produced through this process can thus be used to control water release in pitcher irrigation of tree saplings. Batchelor et al. [1] also indicated similar findings in case of pitcher irrigation and subsurface irrigation using clay pipes. It was found that subsurface irrigation using clay pipes were particularly effective in improving yields, crop quality and water use efficiency (WUE) as well as being cheap, simple and easy to use. Siyal and Skaggs [5] showed through computer simulations with HYDRUS (2D/3D) that with the increase in the pressure head in the irrigation pipe, the size of the wetted zone was also increased. The depth of pipe installation affects the recommended spacing between laterals, although the use of shallow installations can lead to higher evaporative losses. For a given water application, the potential rate of surface evaporation affects the shape of the wetted region only minimally. Soil texture, due to its connection to soil hydraulic conductivity and water retention, has a larger impact on the wetting geometry. In general, greater horizontal spreading occurs in fine-textured soils, or in the case of layered soils in the finer textured layers.

7.2   MATERIALS AND METHODS

Pottery dripper (PD) were designed and manufactured from local materials, where the basic component of PD was pottery discs. It is made from various mixtures, porosity, and volumes. The mixtures contained organic matter (saw dust) by 10, 21 and 31% total mixture weight. The basic element of mixture was Aswan clay. Pottery discs are used in design of polyethylene dripper (Fig. 7.1) to protect and service the pottery discs for basic jobs (dripping and reducing the saline water concentration).

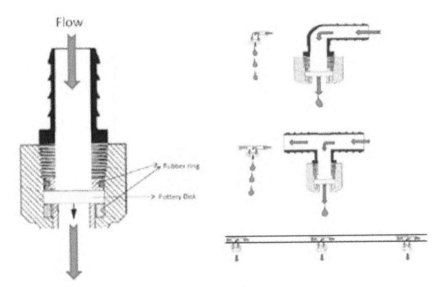

FIGURE 7.1   Schematic diagram of a pottery emitter.

Dimensions of potter emitter were designed for three volumes and three porosities. These six types of pottery drippers were then tested under three pressure heads (0.5, 0.75 and 1 bar) and three concentrations of saline groundwater (6154, 7149 and 7863 ppm). The variables are listed below:

- Pottery disc porosity, according to organic matter (saw-dust) percentage in pottery mixture: 5, 10, and 15% from mixture weight. = P1, P2, P3.
- Pottery disc volumes (disc diameter of 3.5, 2.8 and 2 cm; with thickness of 5 cm): 49, 31.4 and 16.5 cm^3 = V1, V2 and V3
- Operating pressure head: 0.5, 0.75 and 1 bars, and
- Three different water concentrations were used to test pottery drippers test: 6154, 7149 and 7149 ppm = C1, C2, C3.

7.2.1 MEASUREMENTS AND CALCULATIONS

7.2.1.1 HYDRAULIC CHARACTERISTICS

The flow performance of pottery drippers was evaluated under three operating pressure heads (0.5, 0.75 and 1 bar). The emitter discharge was determined by the amount of water application into a graded container during a known time according to Keller and Karmeli [3].

7.2.1.2   EXPERIMENTAL STUDIES

Laboratory experiments were conducted at Hydraulic Laboratory of Desert Research Center in Egypt to test and evaluate the hydraulic characteristics of pottery drippers (PD) with three disc diameters 3.5, 2.8 and 2 cm; and of thickness 5 cm each. The testing was conducted under pressure head range from 5 to 10 m. The laboratory testing equipment (Fig. 7.2) consisted of an electrical centrifugal pump (1 inch/1.5 inch, 25–56 m pressure head, 6.6 m³/h discharge, 1.5 kW, and 2860 rpm), pressure gages (10 cm smallest reading), manometer, control valves, and water container.

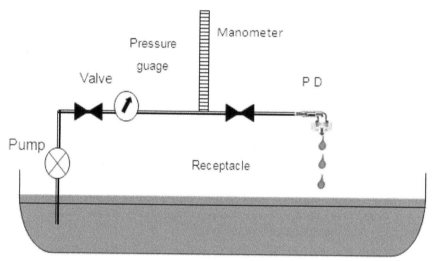

FIGURE 7.2   Laboratory apparatus to test the pottery drippers at Desert Research Center, Egypt.

Hydraulic measurements included: dripper discharge and operating pressure head, measurements for various types of designed nozzles. Application of salinity water was allowed into the emitters. Salinity was measured by electrical conductivity meter, before and after the injection of water through the dripper.

7.2.1.3   PHYSICAL CHARACTERISTICS

Pottery disc surface was photographed by "Optical micrograph and scanning electron microscope (SEM)." Also measurements consisted of determining: water absorption (%), bulk density (g/cm³), drying deflators (%), burning deflators (%), and total deflators (%). The XRD analysis was conducted for three types of pottery mixture at X-ray laboratory besides other measurements at National Center for Housing and Building Research, Egypt.

7.3   RESULTS AND DISCUSSION

7.3.1   REDUCING WATER CONCENTRATION VERSUS POTTERY VOLUME AND POROSITY

Table 1 illustrates that saline water concentration was reduced after using pottery drippers. Saline water with concentration C1 (6154 ppm) was reduced to 5440 ppm. This indicates that concentration was reduced by 714 ppm due to the use of pottery with the smallest porosity percentage (10%). Thereupon, water concentration declined to 5806 ppm and 5840 for pottery porosity of 21 and 31%, respectively.

This is the result of the biggest volume of pottery disc (49 cm³). Volume 31.4 cm³ of pottery disc reduces the water concentration but not as the biggest volume 49 cm³, where water concentration declines from 6154 to 5678, 5678 and 5848 ppm for pottery porosity of 10, 21 and 31%, respectively. However, the smallest volume of pottery discs (16.5 cm³) is the least effective in reducing saline water concentration, where water concentration declines from 6154 ppm to 5780, 5831 and 5848 ppm in pottery disc porosity of 10, 21 and 31%, respectively. These findings indicate that that the decline of saline water concentration is directly proportional to volume of pottery disc, and inversely proportional to pottery porosity percentage (Fig. 7.3).

TABLE 7.1   Effects of Pottery Discs Volume and Porosity on TDS

C3 (7863 ppm)				C2 (7149 ppm)				C1 (6154 ppm)			
31%	21%	10%	P/V	31%	21%	10%	P/V	31%	21%	10%	P/V
7855	7846	7616	V1	6494	6456	6363	V1	5848	5831	5780	V1
7850	7829	7531	V2	6478	6422	6337	V2	5848	5823	5678	V2
7846	7778	7285	V3	6456	6363	6329	V3	5840	5806	5440	V3

P = porosity percentage, %; V = Pottery disc volume, cm³; C = Saline water concentration, ppm; TDS = total dissolved salts.

Saline water with concentration C2 (7149 ppm) is reduced to 6329 ppm. This indicates that concentration shall be reduced by 650 ppm for using pottery with the smallest porosity percentage (10%). Water concentration shall decline to 6363 and 6456 for pottery porosity of 21 and 31%, respectively. This applies to the biggest volume of pottery discs (49 cm³). Volume 31.4 cm³ of pottery disc reduces water concentration from 7149 ppm to 6337, 6422 and 6478 ppm for pottery porosity of 10, 21 and 31%, respectively. However, the smallest volume of pottery disc (16.5

cm³) is the least effective in reducing saline water concentration, where water concentration has reduced from 7149 ppm to 6363, 6556 and 6494 ppm for pottery disc porosity of 10, 21 and 31%, respectively. This clearly demonstrates that reduction of saline water concentration is directly proportional to volumes of pottery disc and inversely proportional to pottery porosity percentage, as shown in Fig. 7.4.

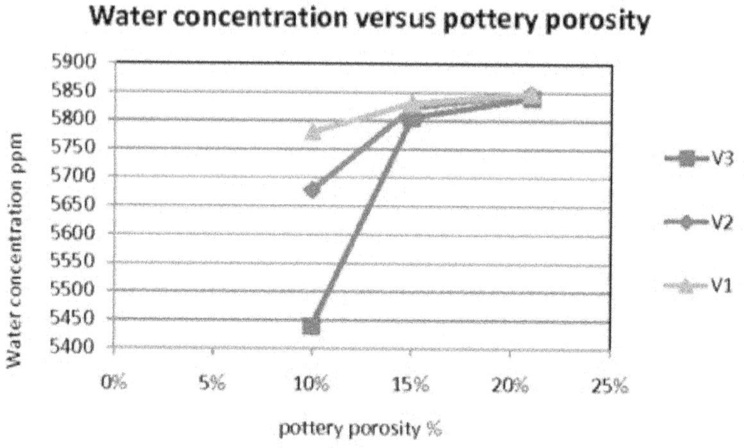

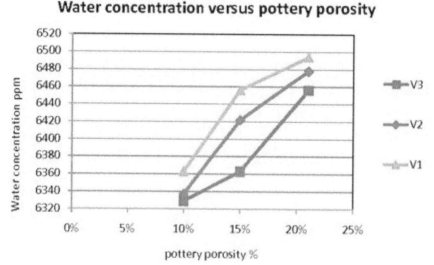

FIGURE 7.3   Effects of dripper porosity on total dissolved salts (6154 ppm) in the water.

FIGURE 7.4   Effect of pottery dripper porosity of total dissolved salts (7149 ppm).

Saline water concentration C3 (7863 ppm) is reduced to 7285 ppm. This indicated that concentration is reduced by 578 ppm for the smallest porosity of 10%.
Thereupon, water concentration declines to 7778 and 7846 ppm for pottery porosity of 21 and 31%, respectively. This applies to the biggest volume of pottery discs
(49 cm³). Volume 31.4 cm³ for pottery disc reduces water concentration from 7863
ppm to 7531, 7829 and 7850 ppm for pottery porosity 10, 21 and 31%, respectively.
However, the smallest volume of pottery disc (16.5 cm³) is the least effective in reducing saline water concentration, where water concentration is reduced from 7863
ppm to 7616, 7846 and 7855 ppm for disc porosity of 10, 21 and 31%, respectively.
This indicates that the reduction of saline water concentration is directly proportional to volume of pottery disc and inversely proportional to pottery porosity percentage, as shown in Fig. 7.5.

Reduction of saline water concentration is attributed to adsorption of Na^+ on
clay minerals that is found in pottery mixtures. Hence, increasing the clay mineral
percentage in pottery mixture shall increase the ability of Na^+ adsorption.

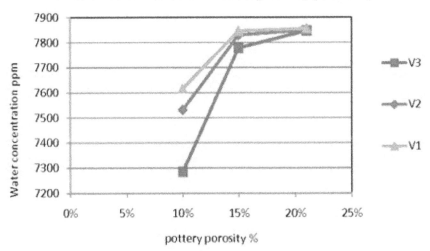

FIGURE 7.5   Effects of pottery dripper porosity on total dissolved salts (7863 ppm).

7.3.2   HYDRAULIC CHARACTERISTICS OF POTTERY DRIPPER

Table 7.2 shows the flow rates versus operating heads, porosity % and disc volume.
Pottery drippers (10% pottery porosity and volume 49 cm³) were tested under operating pressure heads of 0.5, 0.75 and 1 bar. The flow rate was increased from 0.75
and 1.5 to 1.95 lps. In pottery disc volume of 31.4 cm³, the flow rate increased from

0.53, and 0.97 to 1.41 lps. For volume of pottery disc of 16.5 cm³, the flow rate increased from 0.39 and 0.755 to 1.12 lps, as shown in Table 7.2 and Fig. 7.6.

The increase in the flow rate is associated with the increase in the flow area. The flow area is determined according to inside diameter of the rubber ring, where the last flow rates are appropriate for drip irrigation systems.

pottery drippers flow rate under various operating head.

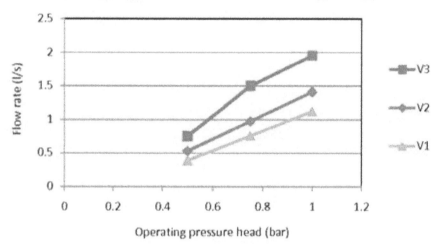

FIGURE 7.6   Effects of operating pressure head on flow rate of pottery dripper (porosity 10%).

TABLE 7.2   Effects of Operating Pressure Head on Pottery Drippers Flow Rate

31%				21%				10%			
49	31.4	16.5	V/H	49	31.4	16.5	V/H	49	31.4	16.5	V/H
19.4	18.2	17.9	0.5	1.34	0.99	0.79	0.5	0.75	0.53	0.39	0.5
28.3	27.9	4.72	0.75	2.53	1.8	1.46	0.75	1.5	0.97	0.755	0.75
38.3	37.7	37	1	3.72	2.51	2.13	1	1.95	1.41	1.12	1

P = porosity percentage, %; V = Pottery disc volume, cm³; H = Operating head pressure, bars.

For 21% of pottery porosity and volume 49 cm³, the flow rate increased from 1.34 and 2.53 to 3.72 lps. In volume 31.4 cm³, the flow rate increased from 0.99 and 1.8 to 2.51 lps. For volume of pottery disc 19.5 cm³, the flow rate increased from 0.79 and 1.46 to 2.13 lps, as shown in Fig. 7.7. The increase in flow rate is associated with the increase in the flow area.

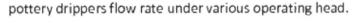

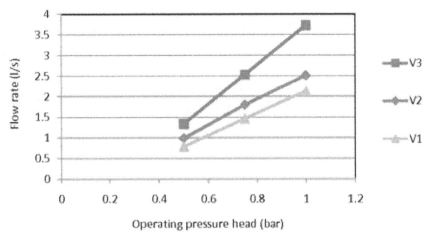

FIGURE 7.7   Effects of pressure head on flow rate of potter dripper (porosity 21%).

For 31% of pottery porosity and volume 49 cm³, the flow rate increased from 19.4 and 28.3 to 38.3 lps. In volume 31.4 cm³, the flow rate has increased from 18.2 and 27.9 to 37.7 lps. For volume of pottery disc 19.6 cm³, the flow rate increased from 17.9 and 27.4 to 37 lps, as shown in Fig. 7.8. The increase in the flow rate is associated with the increase in the flow area. The flow area is determined according to rubber ring inside area, where the last flow rates are appropriate for bubbler irrigation system, and the difference between the three volumes is so weak.

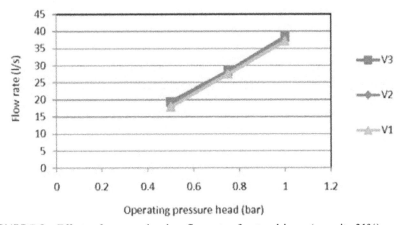

FIGURE 7.8   Effects of pressure head on flow rate of potter dripper (porosity 31%).

It is evident that the increase in the flow rate is associated with the increase in pottery porosity. On the other hand, the efficiency of water salinity reduction is decreased, because of the decrease in the percentage of clay minerals weight in pottery mixture, leading also to a decrease in the cation adsorption process.

7.3.3   PHYSICAL CHARACTERISTICS OF POTTERY EMITTER

Pottery porosity percentage increases with the increase of saw dust in pottery mixtures (Figs. 7.9–7.11). The pottery porosity is 10.47, 21 and 31% for additional sawdust 5, 10 and 15% of mixture weight, respectively. Figures 7.12–7.14 indicate the differences of porosity percentage and random shapes of the pottery discs through use of electronic microscope for pottery disc sections in addition to the optical photo view of the pottery discs that show the quantity of porosity (black sectors) and clay minerals, sand (pink sectors).

Table 7.3 and Fig. 7.15 show that the water adsorption for P1, P2 and P3 is 60, 20 and 31% respectively; and the bulk density is 2.24, 1.7, and 1 g/cm³, respectively.

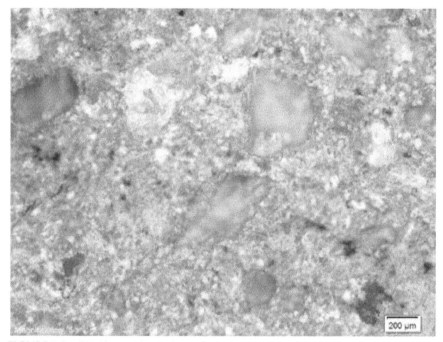

FIGURE 7.9   SEM image shows the surface of pottery disc (porosity 10%).

FIGURE 7.10   SEM image shows the surface of pottery disc (porosity 21%).

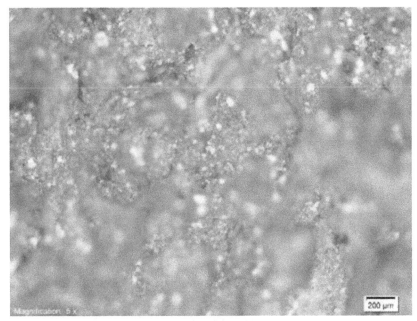

FIGURE 7.11   SEM image shows the surface of pottery disc (porosity 31%).

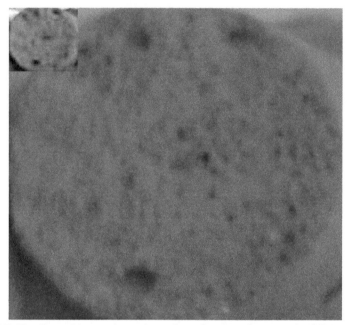

FIGURE 7.12   General image shows the surface of pottery disc (porosity 10%).

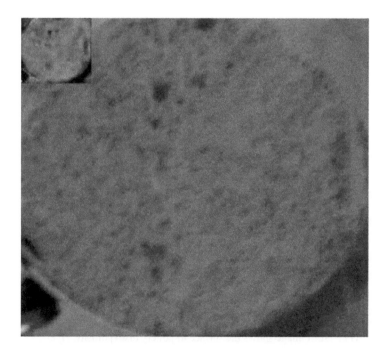

FIGURE 7.13   General image shows the surface of pottery disc (porosity 21%).

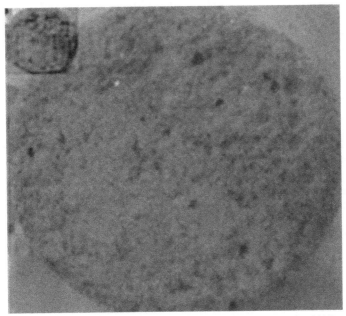

FIGURE 7.14   General image shows the surface of pottery disc (porosity 31%).

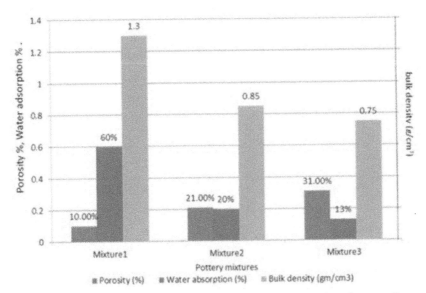

FIGURE 7.15   Effects of organic matter content on physical properties of pottery discs.

TABLE 7.3   Physical Characterizes of Pottery Disc Versus Porosity

Items	Units	P1 (10.47%)	P2 (21%)	P3 (31%)
Water absorption (%)	%	60	20	13
Bulk density (g/cm3)	g/cm³	1.3	0.85	0.75
Drying deflators (%)	%	10.6	10.3	10
Burning deflators (%)	%	1.1	3.8	2.6
Total deflators (%)	%	11.7	17.1	12.6

7.3.4   XRD ANALYSIS FOR THREE TYPES OF POTTERY MIXTURE

Pottery porosity (10%): The mineralogical composition show that the sample consists mainly of quartz, albeit and muttile, moreover illite mineral presets as trace in XRD (Fig. 7.16).

Pottery porosity (21%): From the mineralogical composition of the studied sample, it can be concluded that quartz, muttile and albeit are the primary minerals with no evidence for precedence secondary mineral moreover illite mineral presets as trace in XRD (Fig. 7.17).

Pottery porosity (31%): The mineralogical composition by XRD show that quartz, mullite and muttile are primary minerals representing the main constituents of the studied sample. On the other hand albite were observed as secondary mineral (Fig. 7.18).

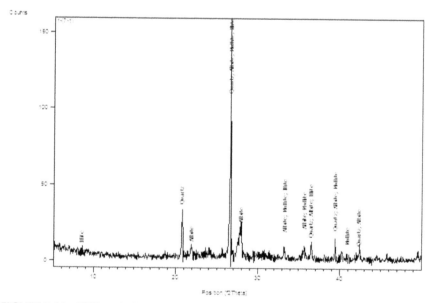

FIGURE 7.16   XRD analysis shows the mineralogical composition of pottery disc (porosity 10%).

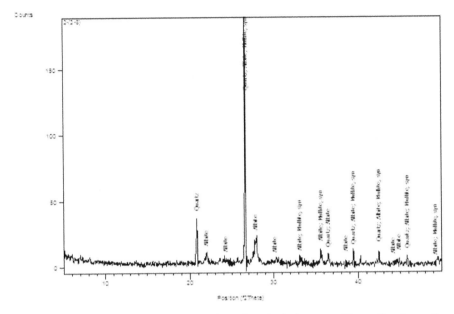

FIGURE 7.17 XRD analysis shows the mineralogical composition of pottery disc (porosity 21%).

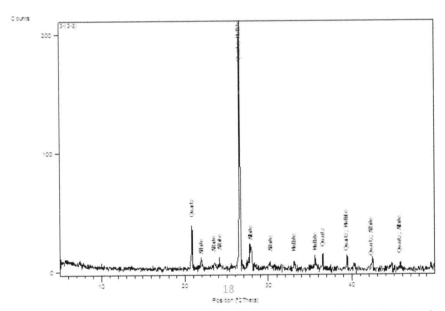

FIGURE 7.18 XRD analysis shows the mineralogical composition of pottery disc (porosity 31%).

7.4 CONCLUSIONS

Pottery dripper (PD) is a new dripper technology for saline water and can reduce water salinity by 750 ppm. The PD is therefore suitable for using saline water and rationalizing the saline water usage in agriculture. This study underscores the advantages of both the pottery media (clay minerals) and irrigation nets to convey water to plants. A new dripper is used as a filter for small water units. Pottery porosity is a basic factor for reduction of water salinity and also flow rate. For porosity of 10% for pottery dripper at operating pressure head of one bar, the discharge ranged from 1.2 to 1.95 lps. For porosity of 21% at operating pressure head one bar, the discharge ranged from 2.13 to 3.72 lps. For porosity of 31% at pressure head one bar, the discharge ranged from 17.9 to 19.4 lps. Saline water concentration decreased from 6154 ppm to 5840, 5806 and 5440 ppm for a porosity of 10, 21 and 31%, respectively. Saline water concentration decreased from 7149 ppm to 7106, 7013 and 6979 ppm for a porosity of 10, 21 and 31%, respectively. Finally saline water concentration decreased from 7863 ppm to 7846, 7778 and 7285 ppm with porosity of 10, 21 and 31%, respectively. It is evident that the increase in the flow rate is associated with the increase in pottery porosity. On the other hand, the efficiency of water salinity reduction was decreased, because of the decrease in the percentage of clay minerals in pottery mixture, leading also to a decrease in the cations adsorption process. Prototypes of pottery drippers were designed and manufactured from local and environmental materials, as one of the models in the agricultural waste recycling industry.

7.5 SUMMARY

Pottery dripper was designed and manufactured to allow the use of saline water in irrigation systems. In Egypt, 2 to 3 thousand million m^3 of saline water are used in the irrigation of about 405,000 ha of land. The main objectives of this study were to: (i) design a new pottery dripper to reduce saline water concentration; and (ii) manufacture and evaluate a pottery dripper. Pottery dripper dimensions were designed for three volumes (49, 31.4 and 16.5 cm^3), three nominal porosities (10, 21 and 31%). These types of PD were tested at three pressure heads (1, 0.75 and 0.5 bars) and three concentrations of saline groundwater (6154, 7149 and 7154 ppm), at Ras-Sudr Research Station in Egypt.

KEYWORDS

- **design**
- **drip irrigation**
- **flow rate**
- **porosity**
- **potter volume**
- **pottery dripper**
- **pressure**
- **saline water**

REFERENCES

1. Batchelor, A. C., Lovella, C., Muratab, M. (1996). Simple microirrigation techniques for improving irrigation efficiency on vegetable gardens. *Agricultural Water Management*, 32(1), 37–40.
2. Bhatt, N., Kanzariya, B., Motiani, A., Pandi, B. (2013). An experimental investigation on pitcher irrigation technique on alkaline soil with saline irrigation water. *International Journal of Engineering Science and Innovative Technology (IJESIT)*.
3. Keller, J., Karmeli, D. (1974). Trickle irrigation design parameters. *ASAE Transactions* 17(4), 678–684.
4. Mathai, M. P., Simon, A. (2002). Water diffusion through pottery discs of varying porosity, *Journal of Tropical Agriculture*, 42(1–2), 63–65.
5. Siyal, A. A. S., Skaggs, T. H. (2009). Measured and simulated soil wetting patterns under porous clay pipe subsurface irrigation. *Agricultural Water Management*, 96, 893–904.
6. Vasudaven, P, Kaphaliy, B., Srivastava, R. K., Tandon, M., Singh, S. N., Sen, P. K. (2011). Buried clay pot irrigation using saline water. *Journal of Scientific and Industrial Research*, 70, 653–655.

CHAPTER 8

ENERGY COST IN DRIP IRRIGATED PEACH ORCHARD

MOHAMED E. EL-HAGAREY, MOHAMMAD N. EL-NESR, HANY M. MEHANNA, and HANI A. MANSOUR

CONTENTS

Modified from Mohamed E. El-Hagarey, Mohammad N. El-Nesr, Hani M. Mehanna, and Hani A. Mansour, "Energy, economic analysis and efficiencies of micro drip irrigation, Parts I and II." *IOSR Journal of Agriculture and Veterinary Science* (IOSR-JAVS), 7(8, August 2014): 10–26. http://iosrjournals.org/iosr-javs/papers/vol7-issue8/Version-2/C07821926.pdf. Modified by the authors. Originally published via open access.

In this Chapter: 1 feddan = 0.42 hectares = 60 × 70 meter = 4200 m² = 1.038 acres = 24 kirat. A feddan (Arabic) is a unit of area. It is used in Egypt, Sudan, and Syria. The feddan is not an SI unit and in Classical Arabic, the word means 'a yoke of oxen': implying the area of ground that can be tilled in a certain time. In Egypt the feddan is the only nonmetric unit, which remained in use following the switch to the metric system. A feddan is divided into 24 kirats (175 m²). In Syria, the feddan ranges from 2295 square meters (m²) to 3443 square meters (m²).

In this Chapter: 1 L.E. = 0.14 US$. The Egyptian pound (Arabic: جنيه مصري‎ Genēh Maṣri; E£ or ج.م; code: EGP) is the currency of Egypt. It is divided into 100 piastres, or ersh [قرش; ʔerš; plural قروش; German: Groschen], or 1,000 millimes [Arabic: mælˈliːm; French: Millime]. The ISO 4217 code is EGP. Locally, the abbreviation LE or L.E., which stands for *livre égyptienne* [French for Egyptian pound] is frequently used. E£ and £E are rarely used. The name Genēh/Geni [geˈneː(h), ˈgeni] is derived from the Guinea coin, which had almost the same value of 100 piastres at the end of the nineteenth century.

8.1   INTRODUCTION

Water resource management is the activity of planning, developing, distributing and managing the optimum use of water resources. It is a subset of water cycle management. Agriculture is the largest user of the world's freshwater resources, consuming 70%. As the world's population rises and consumes more food, industries and urban development's expand, and the emerging bio-fuel crops trade also demands a share of freshwater resources. Water scarcity is becoming an important issue.

This chapter discusses energy, economic analysis and efficiencies of micro drip irrigation to determine the economic impact. The chapter also presents energy consumption and energy indexes in peach production, to investigate the efficiency of energy consumption and to make an economic analysis of peach orchards, according to Zarini and Asadollah [18].

Energy is a fundamental ingredient in the process of economic development, as it provides essential services that maintain economic activity and the quality of human life. Modern agriculture has become very energy-intensive. Energy in agriculture is important in terms of crop production and agro-processing for value adding [7].

Mead [15] defined micro irrigation as slow application of water near the plant. Lubars [14] mentioned following advantages of micro irrigation:

- Optimum growth conditions due to the ability to maintain adequate moisture [9].
- Optimum balance of air, water and nutrients in the soil.
- Better utilization of available space, and plant density can be increased.
- Quicker turnaround of plant materials reducing growth cycles.
- High crop yield.
- Minimize leaching of nutrients that occurs with excess water flow.
- The micro rate system is much cheaper than the common microirrigation systems: smaller P.V.C. tubes size can reduce horse power requirements.
- No runoff on heavy soils.
- No water loss through the root zone on very sandy soils.
- Water and fertilizer saving up to 40–50%.
- Better fruit quality.
- Water can be applied efficiently on shallow soils in hilly areas.

Researchers have compared flow rate from a traditional trickle emitters (8 lph) with that of micro rate emitters (0.4 lph) for the same water quantity of 2.4 L; and they indicated that wetting pattern front in sand and clay soils with traditional trickle flow were faster than wetting pattern front with micro irrigation due to significant water loss deep percolation in a short time. In traditional trickle flow, the vertical wetting pattern fronts increased by 646% in sandy soil than that in clay soils; and the horizontal wetting pattern front in clay soil increased by 8.8% than that in sand in sandy soil [1, 9].

8.2   MATERIALS AND METHODS

During two seasons of 2012 and 2013, research study was carried out in seven years old Florda prince peach orchard (*Purnus perseca L. Batsch*) budded on Nemagard rootstock. The experiment was conducted at the Experimental Farm in Modern Reclamation Lands, situated Bader City, South Al-Tahrir, Al-Beharia Governate, Egypt. Peach trees (seven years) were planted at 5 × 4 m² in sandy soil. This research studied the effects of irrigation using four techniques of drip irrigation systems:

- Gr Surface drip irrigation (SD), 4 lph
- Gr Subsurface drip irrigation (SSD)
- Surface micro drip irrigation (SMD), 0.5 lph
- Subsurface micro drip irrigation (SSMD)

Three amounts of applied water: 60, 80, 100% of calculated applied water = T1, T2 and T3.

For peach trees, amounts of fertilizers are applied according to the recommendations for peach trees by Field Crop Institute, ARC, Egypt, Ministry of Agricultural and Land Reclamation.

Drip irrigation system consisted of a centrifugal pump (5/5 inches with 20 m lift and 80 m³/h discharge), driven by a diesel engine (50 HP), pressure gages, control valves, gate valves, water source from an aquifer, main line, lateral lines and dripper lines. For traditional drip irrigation, Gr dripper (4 lph/m discharge, two dripper at one meter) was used. Each tree row was provided with two hoses to apply 64 lph/tree of irrigation. In micro drip irrigation, one hose for every tree row was used for a discharge of 8 lph/tree using one dripper with a four cross-four distributor to result in 2 lph (Fig. 8.1). In drip irrigation systems, the total dripper discharge for one tree was 64 lph (16 drippers × 4 lph). For micro drip irrigation system, the discharge was 8 lph/tree (4 distributor × 2 lph).

8.2.1   IRRIGATION REQUIREMENTS

Irrigation requirements for peach trees were calculated using local weather station data at Al-Beharia Governorate (Central Laboratory for Agricultural Climate, C.L.A.C.) of Ministry of Agriculture and Land Reclamation. Crop consumptive use (mm/day) was calculated according to Doorenbos and Pruitt [5]. Water requirements for peach trees were calculated with equations by Keller and Karmeli [11]. Results are shown in Tables 8.1 and 8.2.

$$IR = LR + [K_c \times ET_o \times A \times C_F]/[10^7 \times E_a] \qquad (1)$$

where: IR = Irrigation water requirements, m³/ha/day; ET_o = Potential evapotranspiration, mm per day; K_c = Crop factor of peach tree; A = Area irrigated, m²; E_a = Application efficiency = 90% for drip irrigation; LR = Leaching requirements;

and C_F = Coverage factor = 45% for peach trees 45%. The crop factor was used to estimate the ET_{crop} value (mm/day) of peach tree according to FAO [8].

TABLE 8.1 Estimated Consumptive Use (mm/day) of Peach Tree

Growth stage	Month	ETo	Kc	ETcrop	It	Id
		mm/day	–	mm/day	liters/tree/day	m3/ha/day
Initial	January	2.4	0.48	1.152	11.5	5.78
	February	3.2	0.48	1.536	15.4	7.72
	March	4.2	0.48	2.016	20.2	10.11
Mid-season	April	5.6	0.79	4.424	44.2	22.20
	May	6.6	0.79	5.214	52.1	26.17
	June	7.3	0.79	5.767	57.7	28.94
	July	7.2	0.79	5.688	56.9	28.54
End-season	August	6.7	0.75	5.025	50.3	25.21
	September	5.6	0.75	4.2	42.0	21.08
Total, I_y				**5781.44 (m3/ha/season)**		

I_t = Irrigation requirements for one tree per day; I_d = Irrigation requirements for ha per day; I_y = Irrigation requirements for ha per season.

TABLE 8.2 Estimated Irrigation Requirements of Peach Trees for Three Irrigation Regimes

Irrigation treatment	Estimated irrigation requirement (m₃/ha/season)
60% ETC = T1	3468.86
80% ETC = T2	4625.15
100% ETC = T3	5781.44

8.2.2 ENERGY ANALYSIS: MEASUREMENTS AND CALCULATIONS

8.2.2.1 TOTAL ENERGY INPUTS INTO IRRIGATION

The total energy inputs into irrigation were determined on an annual basis and by both area and volume of applied water. The total seasonal energy is the sum of the seasonal fixed installation energy and the seasonal operation energy (pumping plus maintenance) [6]. The seasonal fixed installation energy is the energy required to install the irrigation system for a useful life of at least the length of some evaluation period divided by the number years of the period. In this study, the evaluation period was 20 years. Energy associated with transporting of different components to the site was not considered in this study, because of unreliable data records.

FIGURE 8.1   Micro drip irrigation dripper: 2 lph but every dripper has a cross four distributor to give a 0.5 lph per distributor.

8.2.2.2   PROCEDURES TO CALCULATE THE TOTAL IRRIGATION ENERGY CALCULATIONS

a.   **Installation energy** includes annual fixed energy (AFE) to manufacture a limited number of products used in irrigation system. It was calculated by a method by Batty and Keller [2]:

$$AFE = [(ERM + ERC)(NTR)]/[ESL] \qquad (2)$$

where: AFE = annual fixed energy, MJ/(kg-year); ERM = energy input to manufacture products from raw materials, MJ/kg; ERC = energy input to manufacture products from recycled materials, MJ/kg; NTR = number of times product is replaced over the expected life of the system; and ESL = expected system life, years.

b.   **Energy required manufacturing equipment or machinery**: Batty and Keller [3] mentioned the manufacturing energy (ME) for certain products used in irrigation system. ME, which was used in excavation and land forming, was computed by the following relationship [5]:

$$ME = [(\text{hours on job})/(\text{expected life, h})]$$

$$[(kW \times 14.88 \text{ MJ/kW}) + (\text{Equipment weight, tons} \times 71.2 \text{ MJ/ton})] \qquad (3)$$

where: ME = Manufacturing energy; and kW = Engine power.

c.   **Energy associated with fuel consumption** was computed directly as 41.06 MJ/liter on the basis by Batty et al. [3].

d.   **Energy associated with the repairs and maintenance** of the machinery was estimated as 5 percent of machinery energy inputs by Larson and Fang Meier [12].

e.   **Human labor energy** associated was estimated as follows [10]:

$$EHL = NL \times [CHL/F_c] \qquad (4)$$

where: EHL = human labor energy. MJ per fed; CHL = Energy input coefficient representing the human labor energy = 2.3 MJ/(man-h); NL = Number of laborers required for any operation; and F_c = Field capacity, ha per h.

f.   **Operation energy** includes energy inputs in operation tor irrigation system including maintenance and pumping energies. Annual maintenance energy for irrigation system was roughly estimated as 3 percent of annual installation energy [5]. The pumping energy was calculated directly by the following relationship [5, 13]:

$$PE = k. [(A.D.H.)/(E_p.E_i)] \qquad (5)$$

where: PE = pumping energy, (MJ per fed); k = conversion factor depending on the units used; A = area irrigated, (fed); D = net depth of irrigation water requirement, (m); H = pumping head, (m); E_p = pumping system efficiency; and E_i = irrigation efficiency.

g.   **Human labor energy** associated with labor for system operation and management was determined as follows [5]:

$$EHL = NL.[(t.n.c)/(A)] \qquad (6)$$

where: EHL = human labor energy, (Ml per (ha – yr)); NL = number of laborers required for one irrigation; t = time of one irrigation, (h); n = number of irrigation's in year; c = energy input coefficient representing human labor energy = 1.26 MJ/(man.h); and A = area irrigated, ha. Human labor energy inputs associated with operation and control of the irrigation in this study included manual labor to install water control structures [6]; and it represents an eligible energy input of less than 0.42 MJ /(ha-yr.).

8.2.2.3. ENERGY YIELD

The annual energy yield for crops was calculated according to Rao and Malik [16] and Canakci et al. [4].

Energy intensity (MJ/kg) = [Total energy input, MJ/fed.]/[Peach yield, kg/fed.]

Energy ratio = [Total energy output, MJ/fed.]/[Total energy input, MJ/fed.]

Energy productivity (kg/MJ)=[Peach yield, kg/fed.]/[Total energy input, MJ/fed.]

Net Energy Gain (MJ/ha) = [Total energy output

(MJ/ha) – Total energy input (MJ/Ha)]   (7)

Pumping energy requirements depend on the assumed cultivated area of 50 feddans, and 4 basic control valves. In traditional drip irrigation, the irrigation process was carried out by opening one valve at a time only, and then next valve according to irrigation scheduling. Whereas in ultra-low flow irrigation, all of 8 valves were opened during irrigation process, so that the irrigation operating hours were equal in all of two drip irrigation systems. Energy requirements and energy-applied efficiency (EAE) were determined for various drip irrigation systems according to Batty et. al. [3] as follows:

1. **Power Consumption Use For Pumping Water (Bp):**

$$Bp = [Q \times TDH \times Yw]/[E_i \times E_p \times 1000] \qquad (8)$$

where: Q = total flow volume through the system (m^3); TDH = total dynamic head (m); Yw = water specific weight = 9810 N/m^3; E$_i$ = total system efficiency; E$_p$ = pump efficiency; and 1000 = conversion factor.

2. **Pumping Energy Requirements (Er, kW.h):**

$$Er (kW.h) = Bp \times H \qquad (9)$$

3. **Pumping Energy Applied Efficiency (EAE):**

$$EAE (kg/kW) = [Total fresh yield (kg)]/ [Energy requirements (kW.h)] \qquad (10)$$

where: EAE = pumping energy applied efficiency in kg/kW; and H = irrigation time per season (h).

8.2.3   COST ANALYSIS

8.2.3.1   STORAGE WATER EFFICIENCY (SWE)

Storage water efficiency is the ratio of volume of irrigation water that is beneficially used by the crop to the total volume of irrigation water application:

$$SWE = 100 \times [\text{depth of water beneficially used}]/[\text{depth of applied water}] \quad (11)$$

8.2.3.2. PERCENTAGE OF IRRIGATION WATER SAVING

$$\text{Water saving} = [(I_f - I_n)/I_f] \times 100 \quad (12)$$

where: I_f = water use for control treatment (m³/fed); and I_n = water use for various treatment (m³/fed).

8.2.3.3   WATER USE EFFICIENCY

Water use efficiency (CWUE, kg/m³) is the ratio of crop yield to the consumptive use. In practice, CWUE is expressed as mass of marketable yield per unit volume of water (kg/m³).

$$CWUE = [\text{fruit yield/crop water consumption}] \quad (13)$$

$$\text{Unit production cost, LE/kg} = [\text{Annual irrigation cost, LE/m}^3]/[\text{CWUE, kg/m}^3] \quad (14)$$

8.2.3.4   ECONOMIC EFFICIENCY OF IRRIGATION SYSTEM (EEIS, %)

$$EEIS = (\text{Actual yield, kg per fed})/[\text{typical yield/ kg per ha}] \quad (15)$$

8.2.3.5. ANALYSIS OF ENERGY COST

Micro drip irrigation system was compared with traditional drip irrigation system, and cost of irrigation system was computed according to Worth and Xin [17]. Cost analysis was for an area of 50 feddans. For an area of one feddan, fixed and operation costs were calculated based on market price of 2012 for equipments/materials and operation of irrigation cycle, and for drip irrigation systems. Fixed cost was calculated for equipment and operating irrigation cycle; and for pipes/materials in micro irrigation and drip irrigation systems.
  1.   **Initial cost (IC)**

$$IC, LE/ha = [\text{Price of micro and drip irrigation system, LE}] \times$$
$$[\text{Quantity per ha}] \tag{16}$$

2.   Annual fixed cost (FC, LE/ha/year)

$$FC = D + I + T \tag{17}$$

where: FC = annual fixed cost (LE/year); D = depreciation (LE/year); I = the interest (LE/year); and T = taxes and overhead expenses (LE/year) = assumed as 1.5% from initial cost. Interest on initial was calculated using interest rate per year (assumed as 14% based on local bank rates). Depreciation and interest costs were calculated as follows:

$$D = (IC - DC)/EL \tag{18}$$

$$I = (IC + DC) \times 0.5 \times IR \tag{19}$$

where: IC = initial cost (LE/ha); DC = price after depreciation (LE); EL = expected life (year); and IR = interest rate per year (assumed as 14% based on local bank rates). Capital recovery factor (CRF) was the sum of depreciation and interest on investment, as follows:

$$CRF = [i(1+i)^n - 1]/[(1+i)^n] \tag{20}$$

where: CRF = capital recovery factor; i = the interest rate decimal; and n = the period of analysis in year.

$$\text{Equipment costs per year} = CRF \times \text{initial cost} \tag{21}$$

3.   Annual operating cost (OC)

$$OC = LC + EC + RMC + IS \tag{22}$$

where: OC = operating cost; LC = labor cost (LE/year); EC = energy cost (LE/year); RMC = repair and maintenance cost (LE/year) = RMC was assumed as 3% of initial cost; and IS = installation cost of laterals (LE/year). Labor cost was calculated based on one man for irrigation system. Energy cost was calculated as follows:

$$BHP = Q \times TDH/k \times E \tag{23}$$

where: BHP = break horse power (HP); Q = discharge rate (lph); TDH = total dynamic head (m); k = conversion factor to convert to energy unit = 1.2; and E = the overall efficiency = 55% for pump driven by internal combustion engine. The power cost of diesel type source was calculated as follows:

$$E.C = 1.2 \times BHP \times H \times S \times F.C. \tag{24}$$

where: E.C = energy cost of diesel (LE/HP); H = annual operating hours (h); and S = specific fuel consumption (L/HP.h); F.C. = fuel price (LE); and 1.2 = factor to account for lubrication.

4. **Total annual cost** (LE/year) = FC + OC (25)

5. **Unit production irrigation cost** (LE/kg) =
   [Annual irrigation cost, LE/m³]/[FWUE, kg/m³]   (26)

8.3 RESULTS AND DISCUSSIONS

8.3.1 ENERGY ANALYSIS

Power consumptive use for pumping water (Bp, kW) and energy requirements for SMD (surface micro drip irrigation system) and SSMD (subsurface micro drip irrigation system) irrigation systems were lower than SD (surface drip irrigation system) and SSD (subsurface drip irrigation system) irrigation systems, due to reduction of pumping flow under SMD and SSMD, because micro flow irrigation does not need high capacity of pumping flow. Also the energy requirements of SSMD and SD irrigation systems were 55% lower than those obtained under SD and SSD irrigation systems, as shown in Fig. 8.2. During the two successive seasons, Fig. 8.3 shows energy productivity (kg/MJ) for applied water, and A.I.E.I. (MJ/m³-yr.) for three irrigation regimes (T1, T2, T3) treatments in four (SSMD, SMD, SSD and SD) irrigation systems.

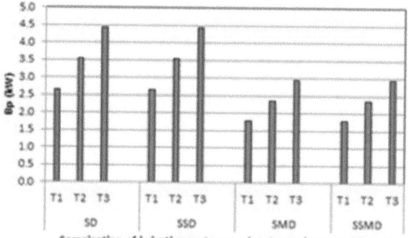

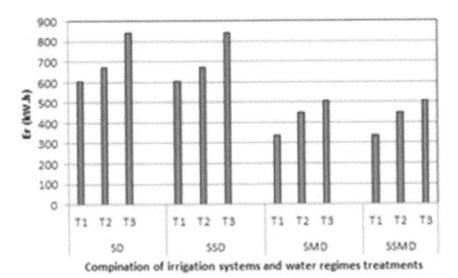

FIGURE 8.2 Power consumption use for pumping water (Bp) and energy requirements (Er) for four irrigation systems in three irrigation regimes.

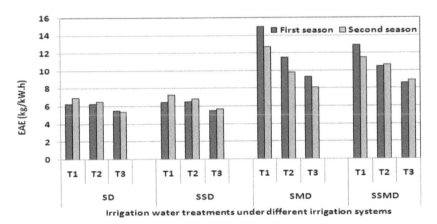

FIGURE 8.3 Energy productivity (kg/MJ) for applied water, and A.I.E.I. (MJ/m³-yr.) for three irrigation regimes (T1, T2, T3) treatments in four (SSMD, SMD, SSD and SD) irrigation systems.

8.3.1.1 INSTALLATION ENERGY

The Fig. 8.4 shows that the installation energy for micro drip irrigation systems (SMD and SSMD) was 13.4% lower than traditional drip systems (SD and SSD).

The average installation energy in SMD and SSMD irrigation systems was 1558 MJ/ha-yr. compared to 1800 MJ/ha-yr. in SD and SSD irrigation systems. Saving of installation energy in SMD and SSMD irrigation systems was due to saving of installation energy elements (annual fixed energy, AFE in MJ/kg-yr.), depending on light weights of irrigation systems elements. The weights of polyethylene pipes for SMD and SSMD were 42 kg/ha, compared to 84 kg/ha in SD and SSD irrigation systems. The other irrigation elements had the same trend. Also the required energy for manufacturing equipment or machinery, the required energy for fuel consumption, the required energy for the repairs and maintenance, and human labor energy depend on the quantity and weight of the materials to install irrigations systems.

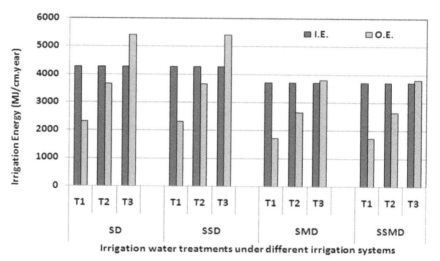

FIGURE 8.4   Installation (I.E.) and operation energies (O.E.) needed for SSMD, SMD, SSD and SD irrigation systems under three irrigation treatments.

8.3.1.2 OPERATIONAL ENERGY

The energy inputs to operate different irrigation systems including maintenance, pumping energies and human labor energy were 970.91, 1540.45 and 2274.42 MJ/ha-yr in SD and SSD irrigation systems under three irrigation regimes (T1, T2 and T3), respectively; and were 729.50, 1108.97 and 1108.97 MJ/ha-yr in SMD and SSMD irrigation systems under T1, T2 and T3 treatments, respectively. The average operational energy inputs in SD and SSD were 28.2% higher than SMD and SSMD irrigation systems. The micro irrigation system saved the inputs of energy operation by 28.2%. Fig. 3 shows that the reduction of energy inputs was obtained by reducing the applied water quantity. Also, there was a saving of energy inputs by using SMD and/or SSMD, because the flow of these two systems was one lph, as four gate valves were opened under the aforementioned irrigation systems to complete the irrigation cycle in a certain time depending on the applied water quantity. On the other hand, only one valve was opened in SD and/or SSD irrigation systems to complete the irrigation cycle to apply the same quantity of water.

8.3.1.3   ENERGY YIELD

8.3.1.3.1   ANNUAL TOTAL IRRIGATION ENERGY INPUTS FOR APPLYING WATER

Annual total irrigation energy inputs for applying water is defined as the total energy inputs of applied water under various irrigation systems (A.I.E.I). It increased by the increasing of the applied water. The highest value of A.I.E.I. was 0.19 MJ/m³-yr in SD and SSD for the highest irrigation treatment, while the lowest value was 0.13 MJ/m³-yr in SMD and SSMD irrigation systems for the same water treatment (Table 8.3).

TABLE 8.3   Energy Applied Efficiencies For Surface Drip (SD), Subsurface Drip (SSD), Surface Micro Drip (SMD) and Subsurface Micro Drip (SSMD) For Three Irrigation Treatments

Energy type	Season	SD			SSD		
		T1	T2	T3	T1	T2	T3
A.T.E.I.	–	6595	7950	9697	6595	7950	9697
A.W.U.	–	34,543	46,058	57,572	34,543	46,058	57,572
A.I.E.I.	–	0.19	0.17	0.17	0.19	0.17	0.17
R.E.C.	1st season	0.879	0.788	0.873	0.708	0.763	0.878
	2nd season	0.788	0.758	0.896	0.631	0.727	0.847
E.E.C.I.	1st season	0.462	0.415	0.46	0.372	0.401	0.462
	2nd season	0.415	0.399	0.472	0.332	0.383	0.446
EP	1st season	1.138	1.269	1.145	1.413	1.311	1.139
	2nd season	1.269	1.319	1.116	1.586	1.376	1.18
NEG.	1st season	2.43	2.65	2.44	2.84	2.71	2.43
	2nd season	2.64	2.71	2.38	3.02	2.78	2.5
		SMD			SSMD		
A.T.E.I.	–	5444	6347	7509	5444	6347	7509
A.W.U.	–	34,543	46,058	57,572	34,543	46,058	57,572
A.I.E.I.	–	0.16	0.14	0.13	0.16	0.14	0.13

Energy type	Season	SD			SSD		
		T1	**T2**	**T3**	**T1**	**T2**	**T3**
R.E.C.	1st season	0.515	0.602	0.526	0.565	0.653	0.452
	2nd season	0.606	0.693	0.59	0.554	0.631	0.534
E.E.C.I.	1st season	0.271	0.317	0.277	0.297	0.344	0.238
	2nd season	0.319	0.365	0.31	0.292	0.332	0.281
EP	1st season	1.94	1.66	1.903	1.771	1.532	2.212
	2nd season	1.652	1.443	1.696	1.805	1.585	1.873
NEG.	1st season	3.3	3.09	3.27	3.18	2.97	3.45
	2nd season	3.07	2.88	3.12	3.21	3.02	3.26

A.T.E.I = Annual total irrigation energy inputs, (MJ/ha-yr);
A.W.U. = Annual water used, (m³/ha-yr);
A.I.E.I = Annual total irrigation energy inputs for applied water, (MJ/m³-yr.);
R.E.C. = Relative consumed energy, (MJ/kg);
E.E.C.I = Energy efficiency of crop irrigation, (%);
EP = Energy productivity, (kg/MJ); and
NEG = Net Energy Gain (MJ/Fed.).

In general, Table 8.3 indicated that the values of the annual irrigation energy input (AIEI) increased by increasing the irrigation water quantity. These values decreased in SMD and/or SSMD. Also, all values of EECI, NEG, and REC had the same trend, except in SD irrigation systems.

8.3.2   ENERGY COST ANALYSIS

8.3.2.1   WATER STORAGE EFFICIENCY (WSE%)

Figure 8.5 shows that the WSE in SSMD was highest compared to that in SMD, because of reduction in losses by deep percolation in sandy soil and evaporation. Also, the actual water amount stored in the effective root zone in SSMD and SMD irrigation systems was higher than that in SSD and SD drip irrigation systems.

8.3.2.2   PERCENTAGE SAVING IN IRRIGATION WATER

During the first season, the highest peach yield was obtained in T2 and T3 treatments under SMD irrigation system. While during the second season, the interaction between the two studied factors indicated that the values were significant using T3 under SSD, and T2 and T3 under SSMD. Therefore, the water saving was 20% using T2 under SMD and SSMD irrigation systems.

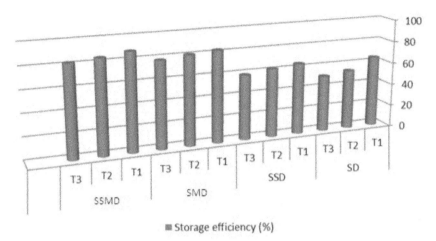

FIGURE 8.5 Water storage efficiency (%), for three irrigation regimes in SSMD, SMD, SSD and SD irrigation systems.

8.3.2.3   CROP WATER USE EFFICIENCY (CWUE, KG/M³)

Figure 8.6 presents crop water use efficiency (CWUE, kg/m³), for three irrigation treatments under SSMD, SMD, SSD and SD irrigation system, during two successive crop seasons. Among three irrigation treatments, T1 gave the highest value of

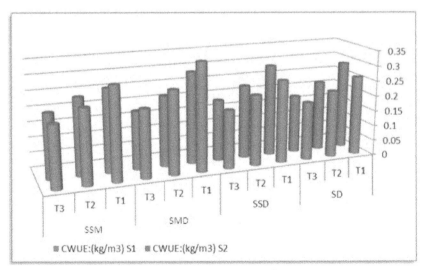

FIGURE 8.6 Crop water use efficiency (CWUE, kg/m³), for three irrigation treatments under SSMD, SMD, SSD and SD irrigation system, during two successive crop seasons.

CWUE compared to T2 and T3 under various drip irrigation systems. The water saving was 40% in T1 compared to 20% in T2, under SSMD and SMD irrigation systems. The differences in values among SD and SSD drip irrigation systems indicated water saving and reduction in nutrients losses by deep percolation and evaporation. These results are in agreement with those Lubars [14]. This gave the peach trees more time to absorb nutrients and water besides having a favorable environment for photosynthesis and respiration processes, which resulted in higher peach yield.

8.3.2.4 IRRIGATION COST PER UNIT PEACH PRODUCTION

Figure 8.7 indicates the irrigation cost per unit irrigation volume (ICIV, LE /m³) for three irrigation treatments (T1, T2 and T3), under SSMD, SMD, and SSD irrigation systems. The ICIV is an economic and important impact factor for farmers and agricultural investors. It considers the irrigation cost of unit weight peach yield, irrigation cost per unit peach production under SSD, SD, SSMD, and SMD irrigation systems, annual fixed cost, annual operating cost and total irrigation cost for three irrigation treatments (Table 8.4). The ICIV was doubled under SSMD and SMD compared to the corresponding values under SSD and SD, due to low capital and annual fixed costs of SSMD and SMD irrigation systems. This is because of smaller pipe sizes in SSMD and SMD irrigation systems.

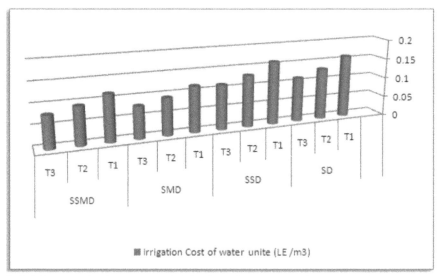

FIGURE 8.7 Irrigation cost per unit irrigation volume (LE/m³), for water applied treatments, under SSMD, SMD, SSD.

8.3.2.5   ECONOMIC EFFICIENCY OF IRRIGATION SYSTEMS

Economic irrigation efficiency (EEIS, %) is an important engineering term that involves understanding of soil and agronomic sciences to achieve the greatest benefit from irrigation. The enhanced understanding of irrigation efficiency can improve the beneficial use of limited and declining water resources, needed to increase and improve crop and food production from irrigated lands. The EEIS is the ratio between actual yield of various water treatments under the studied irrigation systems per feddan. The highest EEIS percentage was achieved under SSMD irrigation system and T3 treatment during the first crop season. Generally, the highest values of EEIS were under SSMD and SMD irrigation systems followed by SSD and SD irrigation systems.

8.3.2.6   COST ANALYSIS

By calculating both annual fixed and operating costs for SSMD, SMD, SSD and SD irrigation systems, it was concluded that SSMD and SMD irrigation systems were more economical than SD and SSD irrigation systems. These differences were due to the increase of capital and annual fixed costs for SSD and SD irrigation systems; and low operating costs due to reduction in repair/maintenance costs, and costs of hoses/pipes; and the reduction in energy requirements. Table 8.4 shows that the capital and annual fixed costs for SSMD and SMD irrigation systems were 25% lower than for SSD and SD irrigation systems. Therefore, the cost of one cubic meter of water in LE for SSMD and SMD irrigation systems was 27–30% lower than SSD

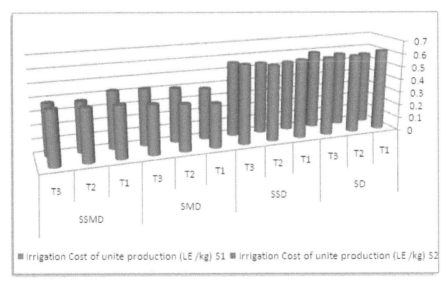

FIGURE 8.8   Irrigation Cost of unit production (LE/kg), for three irrigation treatments, under SSMD, SMD, SSD and SD irrigation system, during two successive growing season.

TABLE 4 Cost Analysis For Surface Drip, Subsurface Drip, Surface Micro Drip and Subsurface Micro Drip Irrigation Systems, For Three Irrigation Treatments

Item	SD			SSD			SMD			SSMD		
	T1	T2	T3	T1	T2	T3	T1	T2	T3	T1	T2	T3
Capital cost for first season, LE/ha ——1												
Control head, A	10,353						10,353					
Drip net cost, B	5,841						2,870					
Hose emitters, C	3,199						1,249					
Subtotal, A+B+C	19,393						14,472					
Annual fixed cost for first season, LE/year ——2												
Depreciation, D	1,454						1,085					
Interest, E	1,697						1,266					
Taxes/Insurance, F	291						217					
Subtotal, D+E+F	3442						2,569					
Annual operating cost, LE/ha ——3												
Labor, G	86						38					
Energy, H	1,432	1,908	2,385	1,431	1,908	2,385	954	1,272	1,590	954	1,272	1,590
Repair/maintenance, K	582						434					
Subtotal, G+H+K	2,100	1,908	2,385	1,431	1,908	2,385	1,426	1,272	1,590	954	1,272	1,590
Annual total cost, LE/ha/season	2,4934	1,908	2,385	1,431	1,908	2,385	1,8467	1,272	1,590	954	1,272	1,590
Irrigation cost per unit water use, LE/m^3	0.16	0.131	0.113	0.16	0.131	0.113	0.116	0.094	0.08	0.116	0.094	0.08

Item	SD			SSD			SMD			SSMD		
	T1	T2	T3	T1	T2	T3	T1	T2	T3	T1	T2	T3
Irrigation treatments												
S1: Irrigation cost per unit production, LE/kg	0.617	0.594	0.594	0.594	0.568	0.594	0.33	0.347	0.366	0.385	0.39	0.402
S2: Irrigation cost per unit production, LE/kg	0.553	0.568	0.594	0.535	0.544	0.564	0.385	0.407	0.423	0.428	0.375	0.383
Capital cost for second season, LE/ha ——— 1												
Control head				4,350					4,350			
Drip net cost				2,454					1,206			
Hose emitters				1,344					525			
Subtotal				8,148					6,081			
Annual fixed cost for second season, LE/year ——— 2												
Depreciation				611					456			
Interest				713					532			

S1 = First season; S2 = Second season.

and SD irrigation systems. Finally, the cost of unit production (LE/kg) for SSMD and SMD irrigation systems was lower than under SSD and SD irrigation systems by 32% and 38.3% in the first season and by 28.7 and 32% in the second season, respectively (Table 8.3, and Figs. 8.7 and 8.8).

8.4   CONCLUSIONS

SSMD and SMD irrigation systems were more economical comparing to SD and SSD irrigation systems, because energy applied efficiency of SMD and SSMD irrigation systems was higher than that for SD and SSD irrigation systems. Average net energy gain in SMD and SSMD irrigation systems for three irrigation treatments is higher than the average net energy gain in SD and SSD irrigation systems for different studied water treatments. The unit peach production of irrigation costs was doubled in SSD and SD irrigation systems compared to SSMD and SMD.

Cost of SD and SSD irrigation systems was higher than SMD and SSMD irrigation systems. Therefore, SSMD and SMD irrigation system were more economical compared to SD and SSD irrigation systems. Irrigation cost of unit peach production under SMD and SSMD irrigation system for three irrigation treatments was lower than under SD and SSD irrigation systems; it was doubled under SSD and SD irrigation systems comparing to SSMD and SMD.

8.5   SUMMARY

This research study was conducted during two successive seasons of 2012 and 2013 in seven-year-old peach trees (*Purnus perseca L. Batsch*) budded on Nemagard rootstock. The experiment was conducted at the experimental farm of Modern Reclamation Lands, situated in Bader City, South Al-Tahrir, Al-Beharia Governate, Egypt. Peach trees were planted at 5×4 m^2 in sandy soil, and were irrigated using four techniques of drip irrigation systems: Gr surface drip (SD) 4 lph, Gr subsurface drip (SSD), surface micro drip (SMD) 0.5 lph, and subsurface micro drip (SSMD) under three irrigation regimes (60, 80, 100% of applied water = T1, T2 and T3). Forty-two trees were selected of normal growth with uniform vigor. Statistical design was split-plot with three replications.

The water saving was 20% for T2 in all of SMD and SSMD irrigation systems. On the other hand, T1 in SMD irrigation systems during the first and second year caused irrigation water saving of 40%. The T1 water treatment gave higher value of CWUE than T2 and T3 in various drip irrigation systems.

The mean value of EEIS was higher in SSMD and SMD irrigation systems than the corresponding value in SSD and SD irrigation systems. Energy applied efficiency of SMD and SSMD irrigation systems are higher than the value in SD and SSD irrigation systems. The higher value of A.I.E.I. is 0.19 MJ/m^3-yr in SD and SSD for T3 water treatment, while the smallest value of A.I.E.I. is 0.13 MJ/m^3-yr in SMD and SSMD irrigation systems for T3 water treatment. A.I.E.I is increasing with applied water increasing. The average energy production in SMD and SSMD irrigation systems is 18.8% higher than SD and SSD irrigation systems. The cost of unit production unit (LE/kg) for SSMD and SMD irrigation systems are 32–38.3% lower than SSD and SD irrigation systems in first season; and 28.7–32% in second season.

Energy applied efficiency in SMD and SSMD irrigation systems are higher than in SD and SSD irrigation systems, as a result of variation of energy requirements and productivity of SMD and SSMD irrigation systems. The SSMD and SMD irrigation systems were more economical compared to SD and SSD irrigation systems. The T2 saved 20% water under SMD and SSMD irrigation systems. The T1 under SMD irrigation system during the first and second year saved 40%. The CWUE was higher in T1 compared to T2 and T3 under various drip irrigation systems. The mean value of EEIS was higher in SSMD and SMD irrigation systems than its counterpart in SSD and SD irrigation systems. The average energy production under SMD and SSMD irrigation systems is 18.8% higher than SD and SSD. The cost of unit production unit (LE/kg) for SSMD and SMD irrigation systems was lower than SSD and SD irrigation systems: 32–38.3% during first season and 28.7–32% during second season.

KEYWORDS

- cost analysis
- crop water requirement
- crop water use efficiency, CWUE
- drip irrigation
- energy analysis
- energy applied efficiency
- energy production
- farm energy
- feddan
- micro irrigation
- peach quality
- peach trees
- subsurface drip irrigation
- surface drip irrigation
- water saving
- water use efficiency, WUE

REFERENCES

1. Abdou, S. H., El-Gindy, A. M. and Sorlini, C. (2010). Performance of ultra-low rate of trickle irrigation. *Misr J. Ag. Eng., Irrigation and Drainage*, 27(2), 549–564.
2. Batty, J. C., Keller, J. (1980). Energy requirement for irrigation. D. Pimentel (ed): *Hand book of Energy Utilization In Agriculture*. Florida, CRC Press. Pages 35–44.
3. Batty, J. C., Hamad, S. N., Keller, J. (1975). Energy inputs to irrigation. *J. of Irri. Drain. Div. (ASC)*, 101(IR4), 293–307.
4. Canakci, M., Topakci, M., Akinci, I., Ozmerzi, A. (2005). Energy use pattern of some field crops and vegetable production: case study for Antalya Region, Turkey. *Energy Convers. Manag.*, 46(4), 655–666.
5. Doorenobs, J., Pruitt, W. O. (1977). *Guidelines for predicting crop water requirements*. FAO Irrigation and Drainage Paper 24. Rome, Italy. Pages 156.
6. Down, M. J., A. K. Turner and T. A. McMahon, 1986. On farm energy used in irrigation. Final Report No. 78/86 of a project supported by the NER. Development and demonstration Council. Melbourne Univ., Civil and Agric. Eng. Dept. 78pp, Australia.
7. Elmesery, A. A. M. (2010). Water movement in soil under micro trickle irrigation system. *Misr J. Ag. Eng., Irrigation and Drainage*, 28(3), 590–612
8. FAO, (1984). *Guidelines for predicting crop water requirements*. FAO Irrigation and Drainage paper No.24.
9. Gilead, G. (2012). Ultra flow drip. http://www.trickle-l.com/new/archives/uldi.html
10. Kassem, A. S. (1986). A mathematical model for determining total energy consumption for agriculture systems. *Misr. J. Agric. Eng.*, 3(1), 39–57.
11. Keller, J., Karmeli, D. (1975). Trickle irrigation design. Rain Bird Sprinkler Manufacturing Corporation, Glendor, CA, 91740 USA. Pages 24–26.
12. Larson, D. L., Fang Meier, D. D. (1978). Energy in irrigated crop production. *Trans. of the ASAE*, 21:1075–1080.
13. Israelson, O. W., Hansen V. E. (Eds.), 1962. Flow of water into and through soils. *Hand book of Irrigation Principal and Practices*. 3rd Edition, John Wiley and Sons, Inc., New York, USA
14. Lubars, P. (2008). http://www.scribd.com/doc/8145273/p13
15. Mead, R. (2002). http://www.americanfarm.com/signe%2010–01.htm
16. Rao, A. R., Malik, R. K. (1982). Methodological considerations of irrigation energetics. *Energy*, 7(10), 855–859.
17. Worth, B., Xin, J. (1983). *Farm mechanization for profit*. Granada Publishing, UK. Pages 250–269.
18. Zarini, R. L., Akram, A. (2014). Energy consumption and economic analysis for peach production in mazandaran province of Iran. *The Experiment*, 20(5), 1427–1435.

CHAPTER 9

PERFORMANCE OF PEACH TREES UNDER ULTRA LOW DRIP IRRIGATION

OMIMA M. EL-SAYED and MOHAMED E. EL-HAGAREY

CONTENTS

Modified from Omima M. El-Sayed and Mohamed E. El-Hagarey, "Evaluation of Ultra-low Drip Irrigation and Relationship between Moisture and Salts in Soil and Peach (*Prunus perssica*) Yield." *Journal of American Science*, 10(8), 12–28, 2014. http://www.jofamericanscience.org/journals/am-sci/am1008/003_24627am100814_12_28.pdf. Modified by the authors. Originally published via open access and used with permission.

9.1 INTRODUCTION

Irrigation affects soil water availability and consequently, plant water status, shoot growth, yield and fruit size [23]. Ultra low drip irrigation (ULDI) uses 10 times less water than common emitters [19]. The advantages of drip irrigation system have been listed in Chapter 8 of this volume [17]. Gilead [12] mentioned that maximal horizontal water distribution under ULDI emitters is lower than the soil infiltration capacity. The lateral and vertical water movement reaches maximum distance from the emission point and is wider than with conventional drip irrigation (CDI). Vertical wetting pattern front in sandy soil increased 36.1% more than in clay. The horizontal wetting pattern front in clay soil increased 13.1% more than horizontal in sand. Abdou et al. [2] concluded that wetting pattern front for sandy and clay soils in CDI was faster than wetting pattern front in ULDI system. In CDI, the vertical wetting pattern fronts in sandy soil increased 646% more than vertical in clay, compared to 8.8% increase in horizontal wetting pattern front in clay soil than horizontal in sand soils.

Peach is considered as one of the most important fruit crops in Egypt with 76,693 feddans producing 332.5 ton in 2011 according to Ministry of Agriculture and Land Reclamation of Egypt.

Abrisqueta et al. [3] found that the continuous deficit and regulated deficit irrigation treatments of Florida star peach showed a lower fruit diameter than the control. Pliakoni and Nanos [24] reported that the deficit irrigation at 50% of ET_c of peach (var. Royal Glory and Caldesi 2000) fruits increased total soluble solid (TSS) and acidity than fruits from control trees. However, Rufat et al. [28] found that irrigation restriction of 28% ET_c during peach growth stage III caused a clear yield reduction in comparison to T_1 (= 100% ET_c) due to direct effects on fruit weight and increasing of TSS and soluble sugar with 30% ET_c. Regulated deficit irrigation (= 35% ET_c) during stage II of peach increased TSS and the TSS/acidity ratio in comparison to control, due to increase in light interception inside canopy tree thus increasing the photosynthetic rate and carbohydrates. Furthermore, decreasing water application based on 25–75% of field capacity significantly decreased average leaf area (cm^2) of almond compared with control, which was irrigated at 100% of field capacity [21]. However, Khattab et al. [14] indicated that increment in irrigation rate was concurrent with increase in chlorophyll a, b and carotenoids in both seasons of pomegranate trees. However, leaf nitrogen and phosphorus content significantly decreased with decreasing the level of water irrigation in almond leaves. While, irrigation with 100% or 75% of field capacity was able to give maximum level of leaf potassium content [21].

This chapter discusses effects of four techniques of drip irrigation systems (Gr surface drip 4 lph, Gr subsurface drip, surface ultra-low drip 1.0 lph, and subsurface ultra-low drip) and three irrigation treatments (60, 80, 100% of calculated applied water called T_1, T_2 and T_3) on yield, fruit quality and leaf parameters of peach trees.

9.2   MATERIALS AND METHODS

Research study was carried out on seven years old Florida prince peach trees (*Purnus perseca* L.) during the two successive crop seasons (2012 and 2013). The experimental set up is detailed in section 8.2 of chapter 8 in this book. Peach trees were planted at 5 × 4 m² in sandy soil. Following treatments were included in the study:

a. Four methods of drip irrigation:
    Gr surface drip (SD) 4 l/h,
    Gr subsurface drip (SSD) with 15 cm depth,
    surface ultra low drip (SUD) 1.0 l/h, and
    subsurface ultra low drip (SSUD) with 15 cm depth.
b. Three irrigation treatments (based on 60, 80, 100% of calculated applied water called T_1, T_2 and T_3) in each method of irrigation.

Seventy-two peach trees of normal growth with uniform in appearance were selected. The experimental design was split plot, where irrigation treatment was in the main plot and irrigation systems were in submain plots with three replication and two trees per replication. The data were statistically analyzed using analysis of variance and Duncan's multiple range test according to Snedecor and Cochran [30]. Fertilizers were applied according to the recommendations for peach crop by Field Crop Institute, ARC, Egypt, Ministry of Agricultural and Land Reclamation.

Soil samples were taken by screw auger at three depths (20, 40, and 60 cm) and three locations along the drip main line. The horizontal and vertical spacing between each sample location was 20 cm. Samples were analyzed for determining both soil moisture and salt accumulation. Results were drawn by SURFER version11 under on color scale for soil moisture 1–50 and for soil salt distribution from 1–100, windows program. The "Kriging" regression method was used for analysis and contour maps. Tables 9.1–9.3 indicate the soil physical properties and chemical analysis of irrigation water that were analyzed [1] at the Central Laboratory, Desert Research Center.

TABLE 9.1   Some Physical Properties of Soil

Soil depth cm	Practical size distribution				FC %	WP %	BD g/cm3
	C. Sand %	F. Sand %	Silt%	Clay %			
0–30	92.8	3.7	2.0	1.5	10	4.8	1.83
30–60	91.5	1.8	0.2	6.5	11	6.3	1.79
60–90	93.1	0.6	0.4	5.9	13	5.5	1.72

TABLE 9.2   Some Chemical Properties of Soil

Soil depth cm	pH	EC ds/m	Soluble Cations, meq/L				Soluble Anions, meq/L			
			Ca^{++}	Mg^{++}	Na^+	K^+	CO_3^-	HCO_3^-	SO_4^-	CL^-
0–30	8.8	2.8	9.1	9.6	8.61	0.69	–	2.34	12.06	13.6
30–60	8.4	0.21	0.82	0.28	0.8	0.2	–	0.73	0.47	0.9
60–90	8.8	0.757	1.8	1.28	3.65	0.84	–	1.47	2.5	3.6

TABLE 9.3   Some Chemical Properties of Irrigation Water

pH	EC ds/m	Soluble Cations, meq/L				Soluble Anions, meq/L			
		Ca^{++}	Mg^{++}	Na^{++}	K^+	CO_3^-	HCO_3^-	SO_4^-	CL^-
6.9	1.63	2.55	1.61	11.9	0.28	-	2.25	2.79	11.3

9.2.1   IRRIGATION REQUIREMENTS

Irrigation system is described in Section 8.2 of Chapter 8 in this book. Irrigation water requirements for peach trees were calculated according to the local weather station data at Al-Beharia Governorate, belonged to the Central Laboratory for Agricultural Climate (C.L.A.C.), Ministry of Agriculture and Land Reclamation. Irrigation amount was based on crop consumptive use (mm/day) according to Doorenbos and Pruitt [7]. Water requirements for peach trees were calculated according to the equations by Keller and Karmeli [13], and these values are listed in Tables 9.1 and 9.2 in Section 8.2 of Chapter 8 in this book. Crop coefficient of peach was used as mentioned in FAO [10]. Water use efficiency (CWUE, kg/m³) was calculated according to Viets [33].

9.2.2   CROP MEASUREMENTS

Following parameters were assessed during the study:
- **Yield and fruit quality measurements**: total yield (kg/tree)
- **Fruit physical characteristics**: A representative sample of 20 mature fruits was taken from each sampling tree to determine average fruit weight (g), volume (cm³), pulp weight (g), fruit length (cm) and width (cm).
- **Fruit chemical characteristics**: Juice total solid percentage (TSS%), treatable juice acidity percentage (as malic acid) and TSS%/ acid ratio were determined according to AOAC [1].
- **Leaf Area (cm²)**: Twenty mature leaves (at third one from the base of the previously tagged nonfruiting shoots from spring cycle) were taken at random in each replication, and were measured by the planimeter in mid June.
- **Leaf total chlorophyll content**: Concentration per unit leaf area was estimated in the field by using SPAD 502 meter (Minolta Co., and Osaka).

- **Leaf NPK content:** In first week of July of both seasons, 20 matures leaves from the middle portion of current year shoots of each replicate were collected to determine macro elements in dry leaf samples. Nitrogen percentage was estimated by micro-Keldahl according to Pregel [25], Phosphorus percentage was determined using atomic absorption spectrophotometer perken Elmer3300 according to Chapman and Pratt [6], and potassium was estimated according to Brown and Lilleland [5].

9.3   RESULTS AND DISCUSSIONS

9.3.1   IRRIGATION WATER SAVING

During 2012, the best growth parameters and highest yield were observed in T_2 and T_3 irrigation treatments under (SUD) irrigation system. Water saving for SUD irrigation system and T_2 was 20%. For 2013, the interaction between the two studied factors proved that (T_3) with the (SSD) and (T_2 and T_3) with (SSUD) had the highest significant values. Therefore, the water saving was 20% in T_2 under all of SUD and SSUD irrigation systems. The T_1 under SUD irrigation system in both successive years gave water saving of 40%. Abrisqueta et al. [4, 10] showed that it is possible to save irrigation water. Rawash et al. [27] studied response of apple trees to water treatments in new reclaimed soils in Egypt. They found that water saving was approximately 50%.

9.3.2   SPATIAL AND TEMPORAL DISTRIBUTION OF SOIL MOISTURE FOR ULTRA-LOW FLOW DRIP AND DRIP IRRIGATION SYSTEMS

The contour maps were plotted for soil moisture distribution under SED and SD irrigation systems (Fig. 9.1). The soil moisture distribution depends on soil texture and structure, field slope, and climate.

Two hours after irrigation: The greatest soil moisture percentage is 38% under SUD irrigation system compared to 11.5% under SD irrigation system. The soil moisture distribution under SUD irrigation system increased gradually at X, Y, and Z axes compared to that in SD irrigation system where it was so sharply distributed in the Y-direction at 50 cm depth.

Ten hours after irrigation: The greatest soil moisture content was 18% under SUD compared to 9.5% under SD. The soil distribution under SUD was more available in the effective root zone of peach trees, where the peach trees effective root zone is 60 cm. However, the mean soil moisture content of SUD was doubled compared to that in SD, specially in effective root zone. The differences are evident in Fig. 9.1.

Eighteen hours after irrigation: Soil moisture content in SUD irrigation system was greater and doubled compared to that in SD. However, the moisture contents were distributed more gradually at X, Y, and Z directions than SD (Fig. 9.1). Distribution of SUD moisture contents was more ideal than SD at X, Y, and Z directions. These findings are in agreement with the results by Abdou et al. [2] and Elmesery [8].

Effective root zone of peach trees is about 60 cm. Therefore, every soil moisture content after 60 cm depth below soil surface is considered deep percolation and as water loss. We can observe that the greatest values of soil moisture content were recorded at the depths 40–60 and 60–80 cm, specially at 10 and 18 h after irrigation. This increased the deep percolation resulting in increased water loss and reduction in irrigation system efficiency.

9.3.3 SPATIAL AND TEMPORAL DISTRIBUTION OF SOIL MOISTURE FOR SUBSURFACE ULTRA-LOW FLOW DRIP AND SUBSURFACE DRIP IRRIGATION SYSTEMS

"One of the main advantages of SSUD and SSD over other irrigation methods is that it has the potential to be most efficient irrigation systems available today. The word potential is stressed because irrigation efficiency not only depends on the irrigation system itself, but also on its proper design, installation and management. Only when designed, installed and managed correctly, it can be more efficient than any other irrigation systems. Figure 9.2 shows contour maps for soil moisture distribution before the start of irrigation.

Two hours after irrigation: It can be noted that soil moisture percentage under SSUD and SSD irrigation systems was similar at the highest content value. However, soil moisture distribution under SSUD irrigation system was increased gradually at X, Y, and Z axes than the soil moisture distribution in SSD irrigation system where it was sharply distributed at in the Y-direction at 50 cm depth (Fig. 9.2).

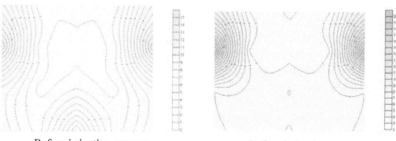

Before irrigation process   Before irrigation process

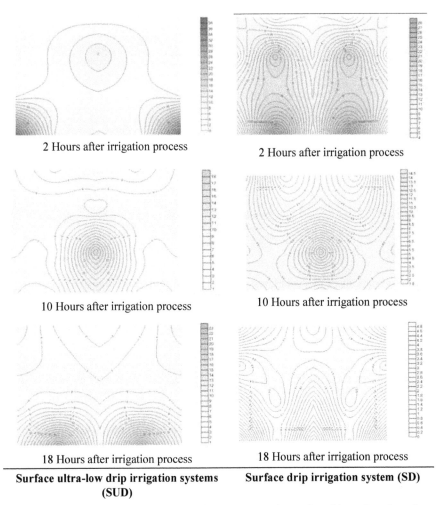

2 Hours after irrigation process

2 Hours after irrigation process

10 Hours after irrigation process

10 Hours after irrigation process

18 Hours after irrigation process

18 Hours after irrigation process

Surface ultra-low drip irrigation systems (SUD)

Surface drip irrigation system (SD)

FIGURE 9.1   The soil moisture distribution patterns under surface drip and surface ultra-low drip irrigation systems.

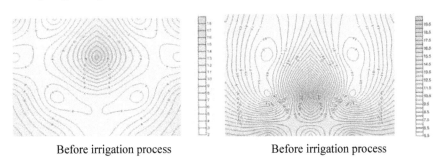

Before irrigation process

Before irrigation process

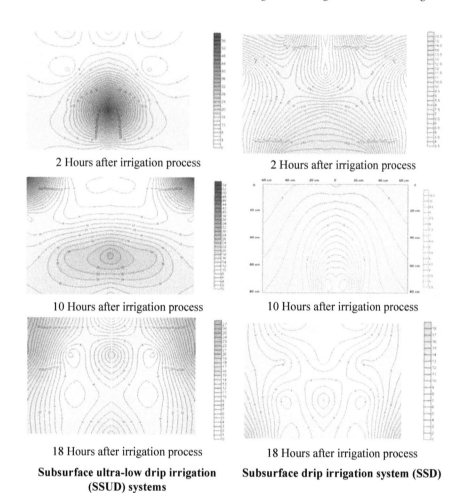

2 Hours after irrigation process   2 Hours after irrigation process

10 Hours after irrigation process   10 Hours after irrigation process

18 Hours after irrigation process   18 Hours after irrigation process

Subsurface ultra-low drip irrigation   **Subsurface drip irrigation system (SSD)**
(SSUD) systems

FIGURE 9.2   The soil moisture distribution patterns under subsurface drip and subsurface ultra-low drip irrigation systems.

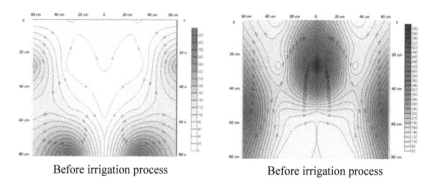

Before irrigation process   Before irrigation process

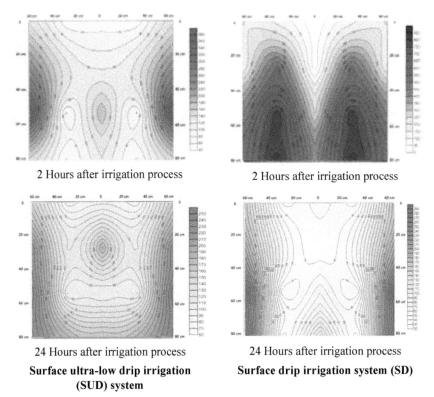

2 Hours after irrigation process   2 Hours after irrigation process

24 Hours after irrigation process   24 Hours after irrigation process

Surface ultra-low drip irrigation   **Surface drip irrigation system (SD)**
(SUD) system

FIGURE 9.3   The patterns of soil salt concentration distribution (EC values) under surface drip and surface ultra-low drip irrigation systems.

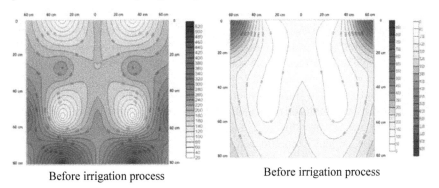

Before irrigation process   Before irrigation process

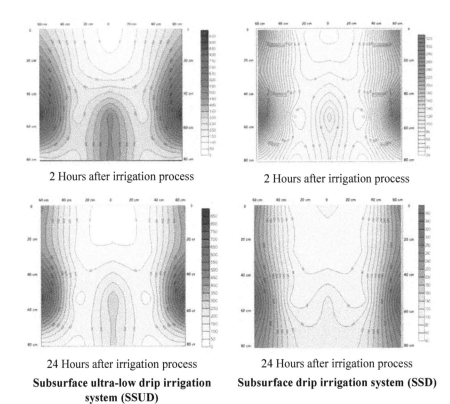

2 Hours after irrigation process   2 Hours after irrigation process

24 Hours after irrigation process   24 Hours after irrigation process
Subsurface ultra-low drip irrigation   Subsurface drip irrigation system (SSD)
system (SSUD)

FIGURE 9.4 The soil salt distribution patterns (EC values) under subsurface drip and subsurface ultra-low drip irrigation systems.

Ten hours after irrigation: The greatest soil moisture content was 55% under SSUD compared to 11% under SSD. The soil moisture distribution under SUD is more available in the effective root zone of peach trees. However, the mean of moisture content of SSUD was almost four times of SSD moisture content specially in effective root zone. The differences are evident in Fig. 9.2.

Eighteen hours after irrigation: Soil moisture contents under SSUD irrigation system were greater than that in SSD, and the moisture content was distributed gradually in X, Y, and Z directions more than in SSD (Fig. 9.2).

Finally, moisture content distribution in SSUD was more ideal than SD at X, Y, and Z directions. These results agreed with those by Abdou et al. [2]. Soil moisture content in SSUD was higher than that in SUD irrigation systems, because the water loss by evaporation under subsurface drip irrigation was less than that of in SUD irrigation system.

9.3.4   SPATIAL AND TEMPORAL DISTRIBUTION OF SALT CONCENTRATION DISTRIBUTION

Under irrigated conditions in arid and semiarid climates, the build-up of salinity in soils is inevitable. The severity and rapidity of build-up depends on a number of interacting factors such as: the amount of dissolved salt in the irrigation water, soil type and local climate. However, with proper management of soil moisture, irrigation system uniformity and efficiency, local drainage, and the right choice of crops, soil salinity can be managed to prolong crop and field productivity. Figures 9.3 and 9.4 show soil salt concentration distribution under SUD and SD irrigation systems.

Two hours after irrigation: It can be observed in Figs. 9.3 and 9.4 that soil salt distribution under SUD, SSUD and SSD irrigation systems was distributed gradually and homogeneously. However, soil salt distribution under SD irrigation system suffered from high concentration at the lower soil layers 40–80 cm, implying that nutrients were exposed to a un-intentional leaching process.

Ten hours after irrigation: The highest salt concentration distribution was at the upper soil layer 0–40 cm under SUD and SSUD irrigation systems, while under SD and SSD irrigation systems the highest salt concentration distribution was at the lower soil layer 40–80 cm (Figs. 9.3 and 9.4).

Due to the high dripper flow (4 lph) compared with SUD flow (1 lph), the vertical movement of water (Y-direction) was more than the horizontal movement of water (X-direction), as a result of water seepage by gravity at in the Y-direction. And the nutrient loss was occurred by deep percolation under effective root zone. For SSUD and SSD, It is logic for the salt concentration of surface soil layers to be higher than the surface soil layer under SD and SUD irrigation systems, because of water evaporation from surface soil which is less under subsurface drip irrigation systems. These results agreed with those by Trooien et al. [32].

9.3.5   THE EFFECTS OF IRRIGATION SYSTEMS, IRRIGATION REGIMES AND THEIR INTERACTION ON YIELD (KG/TREE) AND FRUIT QUALITY OF FLORIDA PRINCE PEACH TREES IN 2012 AND 2013 SEASONS

9.3.5.1   YIELD (KG/TREE)

During the season 2012, Table 9.4 reveals that the SUD had the highest significant fruit yield compared to other irrigation systems. Among the irrigation treatments, T_2 and T_3 gave the higher significant values than the T1. The interaction between the two studied factors revealed that treatment of SUD with T_2 and T_3 had the highest significant value. In the second season, T_3 gave the highest yield. The interaction between the two studied factors proved that SSUD with T_2 and T_3 had higher significant values than most of other treatments.

TABLE 9.4 Effects of Irrigation Systems, Irrigation Regimes (T1, T2, T3) and Their Interaction on Yield (kg/tree) of Florida Prince Peach Trees, During 2012 and 2013 Seasons

Param-eters	Yield (kg/tree)							
season	2012				2013			
WT IS	T_1	T_2	T_3	Mean	T_1	T_2	T_3	Mean
SD	17.98f	20.18de	22.23bcd	20.13C	20.08ab	21.00ab	21.65ab	20.91A
SSD	18.65ef	20.87cd	22.08bcd	20.53C	20.93ab	21.88ab	22.75a	20.86A
SUD	24.10ab	24.68a	24.97a	24.58A	20.40ab	20.97ab	21.68ab	21.02A
SSUD	20.72cd	22.43bc	23.03ab	22.06B	18.53b	22.97a	23.93a	21.81A
Mean	20.36B	22.04A	23.08A		19.99B	21.70AB	22.50A	—

WT = Water treatment; IS = Irrigation system.
Means having the same letter (s) in each column, row and interaction are not significant at 5% level.

TABLE 9.5 Effects of Irrigation Systems, Three Irrigation Regimes and Their Interaction on Peach Fruit Weight (g) of Florida Prince Peach, During 2012 and 2013 Seasons

Items	Fruit weight (g per fruit)							
season	2012				2013			
WT IS	T_1	T_2	T_3	Mean	T_1	T_2	T_3	Mean
SD	58.85abc	62.94a	58.00abc	59.93A	67.04c	74.45bc	85.38ab	75.62AB
SSD	51.24bcd	48.83cd	47.14d	49.07B	74.12bc	73.28bc	91.78a	79.73A
SUD	60.30ab	58.65abc	63.56a	60.84A	71.72bc	70.30c	69.80c	70.61B
SSUD	50.82bcd	59.73abc	63.57a	58.04A	78.14abc	69.53c	75.27bc	74.31AB
Mean	55.31A	57.54A	58.07A		72.75A	71.89A	80.56A	

WT = Water treatments; IS = Irrigation systems
Means having the same letter (s) in each column, row and interaction are not significant at 5% level.

TABLE 9.6 Effect of Irrigation Systems, Three Irrigation Regimes and Their Interaction on Fruit Volume (ml³) of Florida Prince Peach, During 2012 and 2013 Seasons

Items	Fruit volume (ml³)							
season	2012				2013			
WT IS	T_1	T_2	T_3	Mean	T_1	T_2	T_3	Mean
SD	56.90ab	60.30a	55.73ab	57.98A	64.37d	71.30bcd	81.70ab	72.46A
SSD	48.03cd	45.83d	42.17d	45.34B	70.83bcd	68.77bcd	88.43a	76.01A
SUD	56.17ab	55.00abc	57.50ab	56.22A	68.43cd	66.60cd	77.77abc	70.93A
SSUD	53.03bc	55.67ab	58.87ab	55.86A	75.67bcd	66.80cd	73.03bcd	71.83A
Mean	53.53A	54.45A	53.57A		69.82B\	68.37B\	80.23A	

WT = Water treatments; IS = Irrigation systems.
Means having the same letter (s) in each column, row and interaction are not significant at 5% level.

9.2.5.2 FRUIT WEIGHT (G/FRUIT)

During 2012 season, Table 9.5 revealed that the SSD recorded the lowest significant value among all three irrigation treatments. However, insignificant differences were observed among irrigation treatments. The interaction between the two studied factors proved that SUD) with T_3 and SSUD gave higher significant values than most of other treatments. During 20113, among irrigation systems the SSD gave highest significant value. Among water treatments, insignificant differences were noticed among the treatments. The interaction between the two studied factors revealed that SSD with T_3 gave higher significant value than all treatments.

9.2.5.3 FRUIT VOLUME (M³ PER FRUIT)

During 2012 season, Table 9.6 shows that SSD recorded the lowest significant value. Among water treatments, insignificant values were noticed. The interaction between the two studied factors revealed that SD with T_2 had the highest value. In 2013 season, the irrigation systems showed insignificant differences among treatments. Among water treatments, T_3 gave the highest significant value. The interaction between the two studied factors revealed that treatment SSD with T_3 had higher significant value than most of other treatments. These results are in agreement with those by Rufat et al. [28], who reported that deficit irrigation during stage III reduced fruit size and weight of peach fruit that are major attributes to fruit quality. Also, Maria et al. [18] and Singh et al. [26] observed that fruit size, weight and yield values increased with increased amount of irrigation water.

9.2.5.4   FRUIT LENGTH (CM)

During 2012 season, Table 9.7 shows that SD recorded higher significant value. Among three irrigation treatments, T_2 gave the highest significant value. The interaction between the two studied factors proved SD) with T_2 had higher significant value than some of the other treatments. In 2013 season, among all three water regimes and irrigation systems, amounts of water applied and the interaction between the two studied factors were not significantly different among them.

TABLE 7. Effects of Irrigation Systems, Three Irrigation Treatments and Their Interaction On Fruit Length (Cm) of Florida Prince Peach, During 2012 and 2013 Seasons

Items	Fruit length (cm)							
season	2012				2013			
WT / IS	T_1	T_2	T_3	Mean	T_1	T_2	T_3	Mean
SD	4.27abc	4.53a	4.43ab	4.41A	4.97a	5.10a	5.37a	5.14A
SSD	4.03c	4.23abc	4.13bc	4.13C	5.17a	5.17a	5.63a	5.32A
SUD	4.10bc	4.30abc	4.20abc	4.20BC	5.10a	5.00a	4.93a	5.01A
SSUD	4.20abc	4.40abc	4.43ab	4.34AB	5.50a	5.03a	5.10a	5.21A
Mean	4.15B\	4.37A	4.30AB\		5.18A	5.08A	5.26A	

WT = Water treatments; **IS** = Irrigation systems.
Means having the same letter (s) in each column, row and interaction are not significant at 5% level.

TABLE 9.8   Effects of Irrigation Systems, Three Irrigation Treatments and Their Interaction On Fruit Width (Cm) of Florida Prince Peach, During 2012 and 2013 Seasons

Items	Fruit width (cm)							
season	2012				2013			
WT / IS	T_1	T_2	T_3	Mean	T_1	T_2	T_3	Mean
SD	4.90a-d	4.97ab	4.97ab	4.94A	5.07a	5.30a	5.60a	5.32A
SSD	4.57d	4.60cd	4.57d	4.58B	5.13a	5.07a	5.60a	5.27A
SUD	4.70bcd	4.80a-d	5.07a	4.86A	5.27a	5.23a	5.47a	5.32A
SSUD	4.73a-d	4.93abc	4.97ab	4.88A	5.27a	5.37a	5.30a	5.31A
Mean	4.73A	4.83A	4.89A		5.18A	5.24A	5.49A	

WT = Water treatments; **IS** = Irrigation systems.
Means having the same letter (s) in each column, row and interaction are not significant at 5% level.

9.2.5.5   FRUIT WIDTH (CM)

During 2012 season, Table 9.8 shows that the SSD recorded lowest significant value. Among three irrigation regimes, insignificant differences were noticed. The interaction between the two studied factors revealed that SUD) with T_3 had higher significant value than some of the other treatments. In the second season, among all irrigation systems and three irrigation amounts and the interaction between the two studied factors, insignificant differences were noticed. These results are in agreement with Layne et al. [16], who reported that drought conditions negatively impacted tree fruit yield and led to substantial increase in un-marketable fruits. Also, the continuous deficit and regulated deficit water treatments in Florida star peach showed lower fruit diameter than the control [3].

9.3.6   EFFECTS OF IRRIGATION SYSTEMS, IRRIGATION REGIMES AND THEIR INTERACTION ON CHEMICAL CHARACTERISTICS OF FLORIDA PRINCE PEACH TREES IN 2012 AND 2013 SEASONS

9.3.6.1   TOTAL SOLUBLE SOLIDS (TSS%)

During 2012 season, Table 9.9 indicates that TSS% was insignificantly affected by irrigation systems. The differences were not significant among all three irrigation treatments. The interaction between the two studied factors showed a constant trend among treatments. In the second season, the (SSUD) had the highest significant value among all irrigation systems, . Among water treatments, insignificant differences were recorded. The interaction between the two studied factors proved that SSUD with T_3 had the higher significant value. Similar results were reported by Pliakoni and Nanos [24] for Royal Glory peach and Caldesi 2000 nectarine, and Mercier et al. [20] in peach trees (cv. Alexandra). Also, deficit irrigation (30% of ET_c) during stage III [28] and (35% ET_c) during stage II of fruit developing of peach trees [31] increased the total soluble solids.

9.3.6.2   TOTAL ACIDITY, %

In both seasons, Table 9.10 indicated that the SD had highest significant among all irrigation systems. Among three water treatments, T_1 gave the highest value. The interaction between the two studied factors revealed that treatment SD with T_1, T_2 and T_3 in the first season, and T_1 in second season had the highest significant values.

TABLE 9.9 Effects of Irrigation Systems, Three Irrigation Regimes and Their Interaction on TSS% of Florida Prince Peach Fruit, During 2012 and 2013 Seasons

Items								
season		2012				2013		
WT IS	T_1	T_2	T_3	Mean	T_1	T_2	T_3	Mean
SD	12.10a	11.80a	11.00ab	11.63A	12.00bc	12.67abc	11.67bc	12.11B
SSD	11.33ab	11.77a	11.43ab	11.51A	12.67abc	13.00abc	12.67abc	12.78AB
SUD	12.13a	12.10a	10.23b	11.49A	11.33c	11.67bc	12.67abc	11.89B
SSUD	11.33ab	11.33ab	10.87ab	11.18A	13.33ab	13.00abc	14.33a	13.56A
Mean	11.73A	11.75A	10.88A		12.33A	12.58A	12.83A	

WT = Water treatments; IS = Irrigation systems.
Means having the same letter (s) in each column, row and interaction are not significant at 5% level.

9.3.6.3 TSS/ACID RATIO

In 2012 season, Table 9.11 clearly indicated that the lowest significant value of TSS/acid ratio was recorded in SD. Insignificant differences were observed among three irrigation treatments. The interaction between the two studied factors proved that the SUD) with the T_2 had the higher significant value. In the second season, the SUD had higher significant value than first and second treatments, among all irrigation systems. Among three water treatments, T_3 gave higher significant value than T_1. The interaction between the two studied factors revealed that SUD) with T_3 had the highest significant value than other treatments. These results are in agreement with those found by Pliakoni and Nanos [24], who indicated that deficit irrigation with 50% of ET_c in Royal Glory peach and Caldesi 2000 nectarine trees had higher acidity and total phenols than fruits from control peach trees. On the other hand, regulated deficit irrigation (35%ET_c) during stage II of peach fruit developing increased the TSS/acidity ratio compared to the control.

TABLE 9.10 Effects of Irrigation Systems, Three Irrigation Levels and Their Interaction On Total Acidity (%) of Florida Prince Peach Fruit, During 2012 and 2013 Seasons

Items								
season		2012				2013		
WT IS	T_1	T_2	T_3	Mean	T_1	T_2	T_3	Mean
SD	0.93a	0.98a	0.92a	0.96A	1.06a	0.98ab	0.91bc	0.98A
SSD	0.83b	0.81b	0.78bc	0.81B	0.99ab	0.95bc	0.89c	0.94AB
SUD	0.78bc	0.75bcd	0.71cd	0.74C	0.79d	0.78d	0.79d	0.79C
SSUD	0.81b	0.78bc	0.69d	0.76C	0.94bc	0.92bc	0.93bc	0.93B
Mean	0.85A	0.83AB	0.78B		0.95A	0.91AB	0.88B	

WT = Water treatments; IS = Irrigation systems.
Means having the same letter (s) in each column, row and interaction are not significant at 5% level.

TABLE 9.11 The Effects of Irrigation Systems, Three Irrigation Regimes and Their Interaction on TSS%/Acid Ratio of Florida Prince Peach Fruit, During 2012 and 2013 Seasons

Items				TSS%				
season		2012				2013		
WT IS	T_1	T_2	T_3	Mean	T_1	T_2	T_3	Mean
SD	12.31cde	12.25de	11.96e	12.17B	11.38d	13.00bcd	12.81cd	12.39C
SSD	13.64b-e	14.59abc	14.69ab	14.31A	13.02bcd	13.73a-d	14.17abc	13.64BC
SUD	15.62a-e	16.21a	14.42a-d	15.42A	14.24abc	15.04abc	16.10a	15.13A
SSUD	14.05a-e	14.53a-d	15.68ab	14.75A	14.18abc	14.07abc	15.46ab	14.57AB
Mean	13.90A	14.39A	14.19A		13.21B	13.69AB	14.63A	

WT = Water treatments; IS = Irrigation systems.
Means having the same letter (s) in each column, row and interaction are not significant at 5% level.

9.3.6.4 TOTAL SUGARS

In both seasons, the (SUD) had the highest significant value among irrigation systems. Among all water treatments, insignificant differences were noticed in the first season, compared to the highest significant value in T_2 during second season. The interaction between the two studied factors revealed that treatment SUD with T_2 had the highest significant value (Table 9.12). Several reports were in accordance with these results. Gelly et al. [11] reported that an increase in fruit sugar concentration had generally been associated with moderately water stress in peach trees. Also, Kobashi et al. [15] documented an increase in sorbitol, sucrose and total sugars with moderate but not severe water stress. Also deficit irrigation (30% ET_c) during stage III of peach trees increased fruit TSS [28].

9.3.7 THE EFFECTS OF IRRIGATION SYSTEMS, THREE IRRIGATION TREATMENTS AND THEIR INTERACTION ON LEAF CHARACTERISTICS OF PEACH TREES, DURING 2012 AND 2013 SEASONS

9.3.7.1 LEAF AREA (CM²)

During 2012, among irrigation systems the SUD and SSUD had highest significant values (Table 9.13). Among three irrigation treatments, insignificant differences were observed. The interaction between the two studied factors revealed that treatment SUD with T_2 was significant compared to other treatments.

In the second season, among all irrigation systems, the SUD was significantly highest. T_2 and T_3 gave higher leaf area than T_1. The interaction between the two studied factors proved that SUD with T_3 was a highest significant value except SUD) with T_2 treatment. These results are in agreement with the previous finding of Mohy [21], who reported that decreasing irrigation amount from 95% to 25% based on field capacity significantly decreased average leaf area (cm^2) compared to control treatment that was irrigated based on 100% of field capacity.

TABLE 9.12.   The Effects of Irrigation Systems, Three Irrigation Treatments and Their Interaction On total Sugars (GM Glucose/100 mL^3 of Juice) of Florida Prince Peach Trees, During 2012 and 2013 Seasons

Items								
season	**2012**				**2013**			
WT IS	T_1	T_2	T_3	**Mean**	T_1	T_2	T_3	**Mean**
SD	3.43e	3.97c	4.003c	3.80B	3.90f	4.32cd	3.80f	4.02B
SSD	3.85cd	4.09c	3.617de	3.85B	3.91f	4.26cde	3.98ef	4.05B
SUD	4.55b	4.97a	4.417b	4.65A	4.75b	5.22a	4.48bc	4.82A
SSUD	3.92cd	4.03c	3.950cd	3.97B	3.97ef	4.39c	4.03def	4.13B
Mean	3.94A	4.26A	3.997A		4.13B\	4.55A	4.08B	

WT = Water treatments; IS = Irrigation systems.
Means having the same letter (s) in each column, row and interaction are not significant at 5% level.

TABLE 9.13   The Effects of Irrigation Systems, Three Irrigation Levels and Their Interaction On Leaf Area (cm^2) of Florida Prince Peach Trees, During 2012 and 2013 Seasons

Items								
season	**2012**				**2013**			
WT IS	T_1	T_2	T_3	**Mean**	T_1	T_2	T_3	**Mean**
SD	32.24f	33.42ef	34.46c-f	33.37C	33.55f	35.35def	36.14cde	35.01C
SSD	33.87def	34.89b-e	36.06a-d	34.94B	34.32ef	36.10cde	37.09cd	35.84C
SUD	37.05ab	38.08a	37.17ab	37.43A	38.22bc	39.89ab	40.44a	39.52A
SSUD	35.66a-e	36.69abc	37.23ab	36.53A	37.01cd	37.91bc	38.06bc	37.66B
Mean	34.71A	35.77A	36.23A		35.78B	37.31A	37.93A	

WT = Water treatments; IS = Irrigation systems.
Means having the same letter (s) in each column, row and interaction are not significant at 5% level.

9.3.7.2   TOTAL CHLOROPHYLL CONTENT

In 2012 season, Table 9.14 indicates that among irrigation systems SUD gave highest significant value. Among irrigation treatments, T_3 gave a higher significant value than T_1. The interaction between the two studied factors revealed that SUD with T_3 was a highest significant value than most of other treatments. In the second season, among irrigation systems, the SUD gave a highest significant value. Ammon three irrigation treatments, T_2 and T_3 gave higher significant values than T_1. The interaction between the two studied factors shows that SUD) with T_2 and T_3 had higher significant value than most of other treatments. Several past studies were in accordance with the obtained results. Mercier et al. [20] noted that under severe stress photosynthesis and vegetative growth are greatly reduced leading to diminished fruit production. Also, photosynthetic pigment content in leaves was significantly higher in apricot trees grown under high irrigation rates according to El-Seginy et al. [9]. The increment in irrigation rate was concurrent with an increase in chlorophyll-α, β and carotenoids. This increment, in leaf pigments concentration, can be attributed to increased macronutrient uptake especially N and Mg as a consequence of improved soil nutrient status according to Khattab et al. [14].

TABLE 9.14 The Effects of Irrigation Systems, Three Irrigation Levels and Their Interaction on Total Chlorophyll Content of Florida Prince Peach, During 2012 and 2013 Seasons

Items	Total acidity (%)							
season	2012				2013			
WT IS	T_1	T_2	T_3	Mean	T_1	T_2	T_3	Mean
SD	33.46e	34.35de	34.40de	34.07C	36.46 g	37.20 g	37.38fg	37.01C
SSD	35.04cd	35.67cd	35.94bcd	35.55B	38.37ef	39.60cd	39.94bc	39.31B
SUD	36.31abc	37.25ab	37.69a	37.08A	40.86ab	41.30a	41.51a	41.22A
SSUD	35.21cd	35.89bcd	35.84bcd	35.64B	38.61de	39.69bcd	40.00bc	39.44B
Mean	35.01B	35.79AB	35.97A		38.58B	39.45A	39.71A	

WT = Water treatments; IS = Irrigation systems.
Means having the same letter (s) in each column, row and interaction are not significant at 5% level.

9.3.7.3 LEAF MACRO ELEMENT CONTENT, N%

The Table 15 shows that SUD treatment gave highest significant value in both seasons. Among the irrigation levels, T_2 gave highest significant value in the first and second seasons. The interaction between the two studied factors revealed that SUD with the T_2 gave the higher significant value in the first season. However, treatment SUD with T_1 and T_2 gave the higher significant values, during the second season.

TABLE 9.15 The Effects of Irrigation Systems, Three Irrigation Regimes and Their Interaction on Nitrogen Leaf Content (%) of Florida Prince Peach Trees, During 2012 and 2013 Seasons

Items	Total acidity (%)							
season	2012				2013			
WT \ IS	T_1	T_2	T_3	Mean	T_1	T_2	T_3	Mean
SD	2.52cd	2.48d	2.41d	2.47B	2.4c	2.41c	2.24c	2.41C
SSD	2.56bcd	2.69a-d	2.46d	2.57B	2.52c	2.74ab	2.25c	2.59B
SUD	2.83ab	2.9a	2.53cd	2.76A	2.86a	2.95a	2.62bc	2.81A
SSUD	2.460d	2.80abc	2.52cd	2.6B	2.43c	2.81ab	2.49c	2.58B
Mean	2.59B	2.72A	2.48C\		2.55B	2.73A	2.51B	

WT = Water treatments; IS = Irrigation systems.
Means having the same letter (s) in each column, row and interaction are not significant at 5% level.

9.3.7.4   LEAF MACRO ELEMENT CONTENT, P%

In both seasons, Table 16 indicates that SUD treatment had the highest significant value among all irrigation systems. Among all irrigation regimes, T_2 gave the highest significant value. The interaction between the two studied factors showed that SU) with T_2 was the highest significant value.

TABLE 9.16 The Effects of Irrigation Systems, Three Irrigation Regimes and Their Interaction on Phosphorus Leaf Content (%) of Florida Prince Peach Trees, During 2012 and 2013 Seasons

Items	Total acidity (%)							
season	2012				2013			
WT \ IS	T_1	T_2	T_3	Mean	T_1	T_2	T_3	Mean
SD	0.103 g	0.127f	0.103 g	0.111C	0.107 g	0.113 g	0.103 g	0.108D
SSD	0.190d	0.203cd	0.187d	0.193B	0.190ef	0.213cd	0.173f	0.192C
SUD	0.237b	0.267a	0.210c	0.238A	0.243b	0.290a	0.227bc	0.253A
SSUD	0.203cd	0.200cd	0.163e	0.189B	0.197de	0.230bc	0.187ef	0.204B
Mean	0.183B	0.199A	0.166C\		0.184B	0.212A	0.173C\	

WT = Water treatments; IS = Irrigation systems.
Means having the same letter (s) in each column, row and interaction are not significant at 5% level.

9.3.7.5   LEAF MACRO ELEMENT CONTENT, K%

Table 17 indicates that the SUD and SSUD gave highest significant values in the first season. However, in the second season, the SUD gave higher significant value than first and second treatments. Among three irrigation levels in both seasons, T_2 gave the highest significant value. The interaction between the studied factors revealed that SUD with T_2 had the highest significant value.

TABLE 9.17   The Effects of Irrigation Systems, Three Irrigation Regimes and Their Interaction on Potassium Leaf Content (%) of Florida Prince Peach, During 2012 and 2013 Seasons

Items	Total acidity (%)							
season	2012				2013			
WT IS	T_1	T_2	T_3	Mean	T_1	T_2	T_3	Mean
SD	1.10de	1.12cde	1.11e	1.11C	1.11e	1.133de	1.12e	1.12C
SSD	1.167bc	1.187b	1.16bcde	1.17B	1.18cd	1.23bc	1.18cd	1.19B
SUD	1.2b	1.27a	1.163bcd	1.22A	1.23bc	1.29a	1.19bcd	1.24A
SSUD	1.19b	1.22ab	1.207b	1.21A	1.22bc	1.24ab	1.19bcd	1.21AB
Mean	1.17B	1.2A	1.16B		1.18B	1.23A	1.168C\	

WT = Water treatments; IS = Irrigation systems.
Means having the same letter (s) in each column, row and interaction are not significant at 5% level.

These results are in agreement with those by Mohy [21], who found that leaf nitrogen and phosphorus content decreased significantly, with decreasing the level of irrigation amount. While, irrigation with 100% or 75% field capacity resulted in the maximum level of potassium content.

9.3.8   CROP WATER USE EFFICIENCY (CWUE, KG/M³)

Irrigation is an important limiting factor of crop yield, because it is associated with many factors of plant and soil environments, which influence growth and development. Availability of adequate amount of moisture at critical stages of plant growth not only optimizes the metabolic process in plant cells but also increases the effectiveness of the mineral nutrients applied to the crop. Consequently any degree of water stress may produce negative effects on growth and yield of the crop [29]. Surface irrigation method is the most widely used irrigation method all over the world [22].

High variation in CWUE under various water amounts was observed. T_1 irrigation treatment gave higher value than T_2 and T_3 among all three irrigation levels. T1 saved 40$ more water than T2, which saved 20% in SSUD and SUD. The differences were pronounced when compared to SD and SSD drip irrigation systems, due to water saving and reduction in nutrient lost by deep percolation and surface evaporation (Fig. 9.5). Drip irrigation allows more time for the plant to absorb nutrients and water beside favoring an adequate environment for the process of photosynthesis and respiration which reflects positively on crop yield [17].

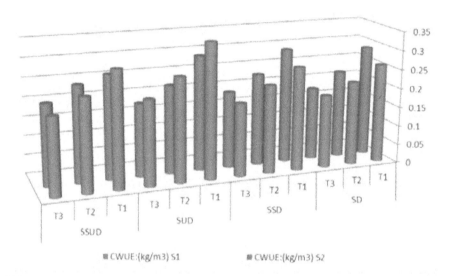

FIGURE 9.5   Crop water use efficiency (CWUE, kg/m³), for three irrigation treatments (T1, T2, T3), under SSUD, SUD, SSD and SD irrigation systems, during the two crop seasons (S1 = 2012, S2 = 2013).

9.4   CONCLUSIONS

Irrigation efficiency involves understanding soil and agronomic sciences to achieve the greatest benefit from irrigation. The enhanced understanding of irrigation efficiency can improve the beneficial use of limited and declining water resources, which is needed to enhance crop and food production from irrigated lands. Ultra-low flow technologies are important methods of irrigation to water management that can reduce loss by runoff in heavy soils or deep percolation in sandy soils.

There was high variation in CWUE values among three irrigation amounts. T_1 treatment gave higher value than T_2 and T_3. T1 saved 40% more water than T2 (which saved 20% in SSUD and SUD. The difference was pronounced in SUD and

SSUD compared to SD and SSD drip irrigation systems. The 40% water and nutrient in sandy soil can be saved, thus increasing quantity and quality of yield by good management and using ultra-low flow drip irrigation. This also will avoid common problems due to excess irrigation like water table rise, aqua fire pollution by loss of nutrients and chemical additions, nutrients and water loss by deep-percolation, nonideal growth environment to plant due to nonmaintain of air balance, and appearance of soil hardpan.

9.5 SUMMARY

This research study was carried out for two successive seasons 2012 and 2013 on seven years old Florida prince peach trees (*Purnus perseca* L.) budded on Nemagard rootstock. The experiment was conducted at the experimental farm, Modern reclamation lands, Situated at Bader City, South Al-Tahrir, Al-Beharia Governorate, Egypt. Peach trees (seven years) were planted at 5×4 m^2 in sandy soil, to study the effects of four techniques of drip irrigation systems (Gr surface drip (SD) 4 lph., Gr subsurface drip (SSD), surface ultra-low drip (SUD) 1.0 lph, and subsurface ultralow drip (SSUD)) under three irrigation depths (= 60, 80, 100% of calculated applied water called T_1, T_2 and T_3) on yield, fruit quality and leaf parameters of peach trees (cv. Florida Prince).

It was concluded that the amount of applied water for T_2 under SUD irrigation system gave the best values of fruit yield and quality, except fruit volume, fruit length, T.S.S. and total acidity % where the highest significant values where obtained with T_2 under SD irrigation systems. However, yield, fruit weight and T.S.S. recorded the highest significant values in T_3 under SSUD. Moreover, the same treatment (T1) increased leaf area and total chlorophyll contents, and mineral contents (N, P and K), during both seasons.

In the first season, the T_2 and T_3 gave the highest yield under SUD irrigation system, causing 20% irrigation water saving in SUD irrigation systems. However, in second season, the interaction between the two studied factors proved that T_3 with SSD and T_2/T_3 with SSUD had the highest significant values. Therefore, there was 20% water saving for T2 under SUD and SSUD irrigation systems. On the other hand, T_1 saved 40% irrigation water by 40% under SUD irrigation system in both of first and second years.

The greatest value of soil moisture content was concentrated at soil depths of 40–60 and 60–80 cm, at 10 and 18 h after irrigation. The effective root zone for peach was 60 cm. Distribution of SSUD moisture contents was more ideal than SD in X, Y, and Z directions. Soil moisture content of SSUD is higher than SUD irrigation systems due to the water loss by evaporation in subsurface drip irrigation that was less than the SUD irrigation system. The higher salt concentration distribution in 0–40 cm depth was under SUD and SSUD irrigation systems, while the higher salt concentration was distributed at 40- 80 cm depth under SD and SSD irrigation systems.

The SUD gave highest significant value. Among three irrigation treatments, T_2 and T_3 gave highest insignificant value. The interaction between the two studied factors revealed that treatment of T_2 and T_3 with SUD had the highest significant value in the first season. In the second season, yield values were not significantly different among all irrigation systems. Among three irrigation amounts, T_3 gave the highest yield. The interaction between the two studied factors proved that T_3 with the SSD and T_2–T_3) with SSUD gave the highest significant values.

KEYWORDS

- consumptive use, ETc
- crop water use efficiency, CWUE
- deficit irrigation
- drip irrigation
- evapotranspiration, ET
- fruit quality
- moisture distribution
- peach trees
- salt concentration distribution
- salt distribution
- subsurface drip irrigation
- surface drip irrigation
- ultra low flow
- water movement
- water saving
- water use efficiency, WUE
- yield

REFERENCES

1. A. O. A. C. (1995). *Official methods of analysis*. 15th ed., Association of Official Agriculture Chemists Washington D.C., USA.
2. Abdou, S. H, Hegazi, A. M., El-Gindy, A., C. Sorlini, (2010). Performance of ultra-low rate of trickle irrigation. *Misr J. Ag. Eng., Irrigation and Drainage*, 27(2), 549–564

3. Abrisqueta, I., Tapia, L. M., Conejero, W., Sanchez-Toribio, M. I., Abrisqueta, J. M., Vera, J., Ruiz-Sanchez, M. C. (2010). Response of early peach (*Prunus persica*)] trees to deficit irrigation. *(Special Issue: Solutions to the water deficit in agriculture.) Spanish Journal of Agricultural Research*, 8(S2), 30–39.

4. Abrisqueta, J. M., Franco, J. A., Ruiz Canales, A. (2000). Water balance in apricot (*prunus armeniaca L.CV. Bulida*) under drip irrigation. *Acta Horticulture*, 537.

5. Brown, J. O., Lilleland, O. (1966). Rapid determination of potassium and sodium in plant material and soil extracts by Flame photometery. *Proc. Amer. Soc. Hort. Sci.*, 48, 344–346.

6. Chapman, H. D., Pratt, P. F. (1961). *Methods of analysis for soil, plant and water*. Div. of Agric. Sci. Univ. Calif., USA. Pages 309.

7. Doorenobs, J., Pruitt, W. O. (1977). *Guidelines for predicting crop water requirements*. FAO Irrigation and Drainage Paper 24. Rome, Italy. Pages 156.

8. Elmesery, A. A. M. (2011). Water movement in soil under micro trickle irrigation system. *Misr J. Ag. Eng., Irrigation and Drainage*, 28(3), 590–612

9. El-Seginy, A. M. (2006). Response of Canino apricot trees to different irrigation and potassium treatments. *Alex. Sci. Exchange J.*, 27:64–75.

10. FAO, (1984). *Guidelines for predicting crop water requirements*. FAO Irrigation and Drainage paper 24.

11. Gelly, M., Recasens, I., Girona, J., Mata, M., Arbones, A., Rufat, J., Marsal, J. (2004). Effects of stage II and postharvest deficit irrigation on peach quality during maturation and after cold storage. *Journal of the Science of Food and Agriculture*, 84:561–568.

12. Gilead, G. (2012). http://www.trickle-l.com/new/archives/uldi.html

13. Keller, J., Karmeli, D. (1975). *Trickle irrigation design*. Rain Bird sprinkler manufacturing Corporation. Glendor, CA, 91740 USA. Pages 24–26.

14. Khattab, M. M., Shaban, A. E., El-Shrief, A. H., Mohamed, A. S. E. (2011). Growth and productivity of pomegranate trees under different irrigation levels, III: Leaf pigments, proline and mineral content. *Journal of Horticultural Science & Ornamental Plants*, 3(3), 265–269.

15. Kobashi, K., Gemma, H., Iwahore, S. (2000). Abscisic acid content and sugar metabolism of peaches grown under water stress. *Journal of the American Society for Horticultural Science*, 125, 425–428.

16. Layne, D. R., Cox, D. B., Hitzler, E. J. (2002). Peach systems trial: The influence of training system, tree density, rootstock, irrigation and fertility on growth and yield of young trees in South Carolina. *Acta Hort.*, 592.

17. Lubars, P. (2008). http://www.scribd.com/doc/8145273/p13

18. Maria, S., Cornel, D., Viorel, S., Ioana, B., Şcheau Alexandru, S., Cristian, D., Radu, B., Manuel, G., Adrian, V. (2010). The influence of the irrigation on peach yield quality in the conditions from peach tree basin Oradea. *Analele Universităţii din Oradea, Fascicula: Protecţia Mediului*. 15:299–302.

19. Mead, R. (2002). http://www.americanfarm.com/signe%2010–01.htm

20. Mercier, V., Bussi, C., Lescourret, F., Genard, M. (2009). Effects of different irrigation regimes applied during the final stage of rapid growth on an early maturing peach cultivar. *Irrigation Science*, 27(4), 297–306.

21. Mohy, A. A. (2011). Effect of some water treatments on growth and chemical composition of almond transplants. M.Sc. thesis, Fac. Agric, Benha Univ., Egypt, 126 pages.

22. Mustafa, O. S., Mustafa, I. M., Arshad, S. A., Sattar, (2003). Adoption of kostiakof model to determine the soil infiltration for surface irrigation methods under local conditions. *International Journal of Agriculture and Biology*, 1, 40–42.

23. Naor, A. (1999). Water stress and crop level interactions in relation to nectarine yield, fruit size distribution, and water potentials. *J. Amer. Soc. Hort. Sci.*, 124(2), 189–193.

24. Pliakoni, E. D., Nanos, G. D. (2010). Deficit irrigation and reflective mulch effects on peach and nectarine fruit quality and storage ability. *Acta Horticulture*, 877, 215–222.
25. Pregel, F. (1945). *Quantitative organic micro analysis*. 4th Ed. London: J. A. Churchil Ltd., Pages 53.
26. Ranbir, Singh, Bhandari, A. R., Thakur, B. C. (2002). Effect of drip irrigation regimes and plastic mulch on fruit growth and yield of apricot (*Prunus armeniaca*). *Indian Journal of Agricultural Sciences*, 72(6), 355–357.
27. Rawash, M. A. A., Zienab, H., Behairy Maryum, M., Hegazi, A., Mostfa, M. (2000). Response of apple trees to some water treatments in new reclaimed soils. *6th International micro irrigation congress*. Cape Town, South Africa 22–27 October.
28. Rufat, J., Arbones, A., Villar, P., Domingo, X., Pascual, M., Villar, J. M. (2010). Effects of irrigation and nitrogen fertilization on growth, yield and fruit quality parameters of peaches for processing. *Acta Horticulture*, 868:87–94.
29. Saif, U., Maqsood, M., Farooq, M., Hussain, S., Habib, A. (2003). Effect of planting patterns and different irrigation levels on yield and yield component of maize (*Zea mays L.*) *International Journal of Agriculture and Biology*, 1, 64–66.
30. Snedecor, G. A., Cochran, B. (1980.) *Statistical methods*. Oxford and J. B.H Bub Com. State Univ. press, Iowa USA.
31. Sotiropoulos, T., Kalfountzos, D., Aleksiou, I., Kotsopoulos, S., Koutinas, N. (2010). Response of clingstone peach cultivar to regulated deficit irrigation. *Sci. Agric. (Piracicaba, Braz.)*, 67(2), 164–169.
32. Trooien, T. P., Lamm, F. R., Stone, L. R., Alam, M., Rogres, D. H., Clark, G. A., Schlegel, A. J. (2000). Using subsurface drip irrigation with livestock wastewater. Proceedings of the 4th decennial Symposium, ASAE, Phoenix, pages 379–384.
33. Vietes, F. G. (1962). Fertilization and efficient of water. *Advanced in Agronomy*, 14:223–264.

CHAPTER 10

WATER PRODUCTIVITY AND FERTILIZER USE EFFICIENCY OF DRIP IRRIGATED MAIZE

H. M. MEHANNA, M. M. HUSSEIN, and H. ABOU-BAKER NESREEN

CONTENTS

Modified from H. M. Mehanna, M. M. Hussein, and H. Abou-Baker Nesreen, "The Relationship between Water Regimes and Maize Productivity under Drip Irrigation System: A Statistical Model." *Journal of Applied Sciences Research*, 9(6), 3735–3741, 2013. http://www.aensiweb.com/old/jasr/jasr/2013/3735-3741.pdf. Modified by the authors. Originally published via open access.

In this chapter: 1 feddan = 0.42 hectares = 60 × 70 meter = 4200 m2 = 1.038 acres = 24 kirat. A feddan (Arabic) is a unit of area. It is used in Egypt, Sudan, and Syria. The feddan is not an SI unit and in Classical Arabic, the word means 'a yoke of oxen': implying the area of ground that can be tilled in a certain time. In Egypt the feddan is the only nonmetric unit, which remained in use following the switch to the metric system. A feddan is divided into 24 Kirats (175 m²). In Syria, the feddan ranges from 2295 square meters (m²) to 3443 square meters (m²).

10.1   INTRODUCTION

Agriculture uses 71% of freshwater in the world. Therefore, innovations and new technologies are needed to increase the irrigation application and conveyance efficiencies to solve problems related to water scarcity. Pressurized irrigation technologies and new methods of irrigation scheduling can be adapted for more effective and rational use of limited supplies of water [18]. Drip and sprinkler irrigation methods are preferable to less efficient gravity methods of irrigation.

All cultivated land in Egypt has an arid or semiarid climate, and the water required for agricultural and horticultural crops is obtained mainly through irrigation systems, which consume about 83% of the country's available fresh water [12]. On the other hand, field application efficiency in most traditional irrigation methods is still very low, typically less than 50% and often as low as 30% [23]. Excessive application of water generally causes losses due to surface run-off from the field and deep percolation below the root zone. Alternative water application methods such as the drip irrigation method allow for much more uniform distribution as well as more precise control of the amount of water applied and also decrease nutrient leaching [26].

Frequency of water application is one of the most important factors in drip irrigation management because of its effects on soil water regime, root distribution around the emitter, the amount of water uptake by roots, and the amount of water percolation under the root zone [5, 8, 36]. Due to these phenomena of irrigation frequency, water use efficiency (WUE) and crop yields may be different under different irrigation frequencies. Irrigation frequency, which results in either excessive or inadequate water application in each irrigation, can have a negative impact on either drip irrigation efficiency or final grain yield. Very low irrigation frequency may cause crop water stress between irrigations, especially in sandy soils because of shorter duration of water application compared to the time over which plants take up water.

Maize is an important cereal crop grown throughout the world [13]. Maize has high irrigation requirements [28, 34]. In arid and semi arid regions, the daily evapotranspiration rates of maize often exceed 10 mm/day for significant time periods [17]. Furthermore, maize yields are most sensitive to water stress, especially at flowering and pollination stages. Although several studies have shown positive responses in some crops to high-frequency drip irrigation, yet his is not the case for maize in different soils and regions. Furthermore, there are inconsistencies in the available data for optimum frequency. Therefore, it is important to determine a proper drip irrigation frequency that promotes maize yield for specific localities, thereby avoiding water stress or water leaching from the root zone.

The objective of this study was to evaluate the impact of drip irrigation frequency on maize yield and WUE to develop a best management drip irrigation practices for high maize yield and water use efficiency (WUE). Authors developed a statistical model to consider the relationship between water frequencies and maize yield in a semiarid regions.

10.2 MATERIALS AND METHODS

During the 2011 and 2012 summer seasons, this study was conducted at the Experimental Farm of the National Research Center, El-Nubaria, El-Beheira, Egypt. The farm is located at 30°25′55.6″N and 30°19′10.5″E. The weather is hot and dry from May to October where temperatures can reach up to 40 °C. The soil at the experimental site was sandy, and the physical properties of soil were recorded in Table 10.1. The organic matter content was 0.1%. The soils have no salinity and drainage problems.

TABLE 10.1 Physical Properties of Soil at the Experimental Site

Characteristics	Value
pH (1: 2.5 soil: water ratio)	8.20
EC (Soil paste extraction) $dS.m^{-1}$	1.32
Soluble cations (m.e./100 g soil)	
Calcium	0.48
Magnesium	0.12
Potassium	0.69
Sodium	0.06
Soluble anions (m.e./100 g soil)	
Bicarbonate	0.22
Chloride	0.77
Sulphate	0.36
Available nutrients (ppm)	
Nitrogen	29.7
Phosphorus	7.90
Potassium	116.63
Physical properties (%)	
Organic matter	0.47
Calcium carbonate	24.9
Sand	68.91
Silt	16.57
Clay	14.52
Textural class	Sandy loam
Soil Taxonomy	Entisol-Typic Torripsamments

A randomized complete block split plot design with four replications was used in each season. Irrigation systems (subsurface drip at 0.2 m depth and surface drip) were randomly assigned to the main plots, and irrigation water frequencies were assigned to the split plots. The drip irrigation system was divided into three sections. Each section was provided with one check valve, one flow meter, and one pressure gage to control the operating pressure and to measure the irrigation quantity. During the season, the three irrigation frequencies were 2500 m^3/fed. (100% of ETc, control), 2050 m^3/fed./season (80% of ETc), and 1900 m^3/fed. (70% of ETc). The amount of irrigation water was applied according to the daily reference evapotranspiration (ETo) computed using the Modified Penman-Monteith equation [4] and daily climatic data, which were obtained from the Central Laboratory of Agricultural Climate (CLAC) for Nubaria location. Thereafter, the calculated ETo values with the crop coefficient (Kc) were used to calculate the amount of water requirement for maize (mm/fed.) with the following equation:

$$ETc = ETo \times Kc \tag{1}$$

The **crop coefficient** is the ratio of the crop evapotranspiration to the reference evapotranspiration and represents an integration of the effects of selected primary characteristics that distinguish it from the reference crop grass [1]. As recommended by Allen et al., [3, 4] and Neale et al. [25], K_C was adjusted according to local climatic conditions, including minimum relative humidity, wind speed, crop growth stage [15] and maximum plant height. The adjusted monthly Kc values during the maize season varied from 0.35 to 1.30, and were calculated in those periods in which plants were not under water stress. The drip irrigation efficiency was assumed as 0.9, and the root extension coefficient according to Goyal [15] and Moon and Gulik [24].

Water use efficiency (WUE) and irrigation water use efficiency (IWUE) were calculated using equation below [2–4, 15, 20].

$$WUE = (Yield, kg/fed.)/(Irrigation frequency, m^3/fed.) \tag{2}$$

Maize crop was sown with 75 cm furrow spacing @ seed rate of 30 kg.ha^{-1}. The regular tillage and agricultural operations of growing maize were followed. All other agronomic practices were kept normal and uniform in all treatments. Fertilizers were based on the recommendations of the regional extension service of Egypt: 240 kg of N/ha, 36 kg of P$_2$O$_5$/ha and 57 kg of K$_2$O/ha. Representative plant samples were collected after 90 days from sowing. And the growth parameters were recorded for plant height (cm), number of leaves per plant, dry weight of leaves, stem, root and whole plant. After harvest, recorded data were: grain yield, NPK uptakes, and water use efficiency (WUE). Finally the statistical analysis and regression equations were fitted to the data to develop relationships between water quantities and the productivity of maize crop, under semiarid conditions in Egypt.

All data were analyzed using an analysis of variance appropriate for a randomized complete block split plot design. Mean separation of treatment effects in this study was accomplished using Fisher's protected least significant difference (LSD) test. Probability levels lower than 0.05 were categorized as significant. All data analyzes in this study were accomplished using the MSTAT as described by Snedecor and Cochran [31].

10.3 RESULTS AND DISCUSSION

10.3.1 GROWTH PARAMETERS

Table 10.2 shows the effects of irrigation systems (subsurface and surface drip), irrigation levels, and the interaction between these two on the growth parameters of maize crop. It was observed that all growth parameters were significantly different of observed under two irrigation systems and irrigation treatments; except for plant height, number of leaves per plant and dry weight of root. Data indicated that the highest values of maize growth parameters were gained by irrigating plants with 100% of the ETc (= 2500 m³/fed.) treatment in subsurface drip irrigation system. On the contrary, the lowest values were observed in plots irrigated with 1900 m³/fed. (70% of ETc) in surface drip irrigation system. There were not significant differences between the growth parameters values among two irrigation systems, but the differences increased by using the different water quantities. The growth parameters decreased by decreasing water quantity from 100% of ETc to 70% of ETc. For example by using 100% of ETc (control), dry weight of whole plant decreased by 46% and 35% under subsurface drip irrigation system, and by 36% and 27% under surface drip irrigation system comparing to other two water quantities (80% of Etc and 70% of Etc), respectively.

Decreasing irrigation water has a harmful effect on maize growth parameters. Kramer and Boyer [21] reported that the plant growth is controlled by rates of cell division and enlargement and by the supply of organic and inorganic compounds required for the synthesis of new protoplasm and cell walls. Cell enlargement is particularly dependent on at least a minimum degree of cell turgor; and stem and leaf elongations are quickly checked or stopped by water deficits. Many investigators Batanouny et al. [6], Ahmed and Mekki [2], El-Noemani [10], El-Sheikh [11], and Mahrous [22] reported that growth criteria of maize plants were reduced when plants were subjected to drought [9]. The physiological mechanisms involved in cellular and whole plant responses to water stress generate considerable interest and are frequently reviewed by the researchers [9, 30]. Drought stress in plants occurs when evaporative demand exceeds water uptake. Water deficit budgets lead to numerous physiological alterations, both in the long-term and the short-term. Long-term drought responses include altered root to shoot ratio [7] and reduced leaf area [6]. Short-term responses include altered stomatal function [33] and osmotic adjustment

[35]. According to Kramer and Boyer [21], plants respond to drought either by delaying dehydration where the plant maintains relatively high plant water potential or by tolerating dehydration where the plant continues to function at lower plant water potentials. Drought has different effects on maize plants depending on the development growth stage at which it occurs. Previous reports showed that stress during tasseling and silking was most harmful; and stress during grain filling was more drastic than that during the vegetative growth stage [16]. Changes in morphological characters are the ultimate determinants of stress effects on plants [14, 19].

TABLE 10.2 Effects of Irrigation Systems and Water Regimes on Growth Parameters of Maize Crop

Irrigation system	Water regime m³/ fed./season	Plant height, cm	Number of leaves/ plant	Dry weight (g)			
				Leaves	Stems	Roots	Whole plant
Subsurface drip	2500 (control)	155.00	11.33	19.31	20.43	10.11	39.74
	2050	154.67	10.33	17.61	15.44	9.41	33.05
	1900	129.33	9.33	10.76	10.57	6.80	21.33
Mean		146.33	10.33	15.89	15.48	8.77	31.37
Surface drip	2500 (control)	146.00	10.00	15.41	12.87	10.13	28.27
	2050	135.33	9.67	13.48	11.38	9.18	24.86
	1900	124.00	8.67	9.89	8.02	6.72	17.9
Mean		135.11	9.44	12.92	10.76	8.68	23.68
Mean values under the effect of water regimes	2500 (control)	150.50	10.67	17.36	16.45	10.12	·34.01
	2050	145.00	10.00	15.55	13.41	9.29	28.96
	1900	126.67	9.00	10.32	9.29	6.76	19.62
L.S.D. at P = 5%	Irrigation System (I)	N.S.	N.S.	2.36	1.12	N.S.	3.48
	Water regimes (W)	18.58	N.S.	1.72	2.24	1.79	3.31
	I × W	26.28	N.S.	2.44	3.17	0.78	4.68

All values are averages of the two growing seasons. I = Irrigation method; W = Irrigation regimes.

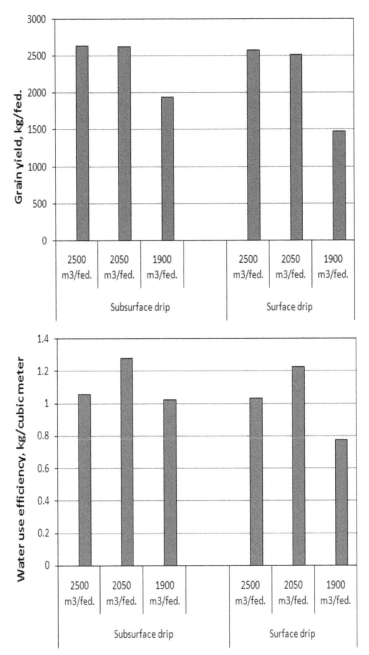

FIGURE 10.1 Effect of the interaction between irrigation system and water regimes on: maize grain yield (top) in kg/fed.; and water use efficiency (WUE, bottom), kg/m³.

10.3.2 YIELD AND WATER USE EFFICIENCY (WUE)

The results of maize grain yield are presented in Fig. 10.1. The highest values were observed under subsurface drip irrigation system for the three irrigation levels (100%, 80% and 70% of ETc = T1, T2 and T3) compared to corresponding values in surface drip irrigation system. There were significant differences between maize grain yield values (2638 and 2575 kg/fed. In subsurface and surface drip irrigation system, respectively) using irrigation water quantity of 2500 m³/fed. (control, T1) compared to the other two water quantities (T2 and T3), respectively. Therefore, water stress affected maize grain yield, especially under surface drip irrigation system. The moderate water quantity (80% of ETc or 2050 m³/fed.) resulted no significant grain yield value compared to the control. There, 20% (app. 450 m³/fed.) of water quantity can be saved.

Data on water use efficiency (WUE) emphasized that the moderate water quantity (2050 m³/fed.) is suitable for the economical maize grain yield, where the highest WUE value (1.28 and 1.22 kg/m³, under subsurface and surface drip irrigation system) was obtained with 80% of ETc water regime, respectively. These results confirm those of Quaranta et al. [27], who found that postsowing irrigation with two further applications gave highest grains/ear and grain yield of maize genotypes. Simpson [29] reported that the variations in yield and its components due to drought stress at different growth stages can be ascribed to the impairment of many metabolic and physiological processes in plants. In this regard, Song et al. [32] showed that water stress led to slower pollen; and filament development decreased the filament fertility and resulted in the reduction of grain number and weight per ear. Similar results were recorded by Ahmed and Mekki [2], Batanouny et al. [6], El-Sheikh [11], and Grant et al. [16].

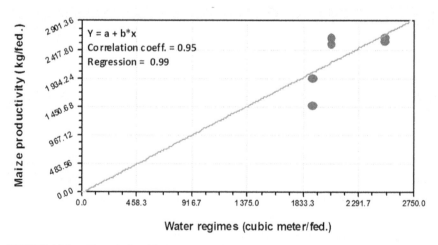

FIGURE 10.2   The relationship between maize productivity and irrigation water quantity.

The relationship between maize grain yield and water quantities is shown in Fig. 10.2 under semiarid conditions in Egypt. The relationship was linear with a coefficient of correlation of 0.95 and followed a straight line: $Y = a + b*X$, where: $Y =$ maize grain yield (kg/fed.) and $\times =$ irrigation water quantity (m³/fed.). The regression coefficients were: $a = -17.097$ and $b = 1.07627$. This statistical model can be used to predict maize grain yield under similar conditions of this study.

10.3.3   UPTAKE OF MACRO NUTRIENTS (N, P AND K PERCENTAGES)

Data for N, P, and K uptake by maize plant plotted in Fig. 10.3. It can be observed that there were no significance differences in N, P and K percentages among irrigation methods and irrigation regimes. Researchers [2] found that P and K concentration in maize grain were increased as the amount of water increased without a significant difference. On the contrary, the significant differences were observed for N, P and K uptakes among three irrigation regimes, and the interaction between water quantities and irrigation systems. However, there were no significance differences among the irrigation methods. Generally, N, P and K uptakes (mg/plant) values were increased by increasing irrigation water quantity. This increment was perhaps attributed to the effect of water as a solvent liquid on both fertilizers and native soil nutrients. In T3 (2500 m³/fed. Based on 100% ETc, control), nitrogen uptake was decreased under surface drip irrigation. This confirms that nitrogen fertilizer was leached easily with addition of high irrigation rate under surface drip irrigation system.

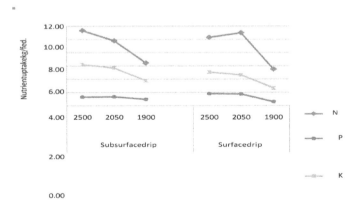

FIGURE 10.3   Effects of irrigation systems and water regimes on N, P, and K uptakes by maize plants (Top curve: N; Center curve: P; and lower most curve: K).

Moreover, the highest values of N, P and K uptakes were 168.35, 20.31 and 920.79 mg/plant in the control irrigation treatment under subsurface drip irrigation system. But the lowest values were 68.57, 6.64 and 326.13 mg/plant using the lowest water quantity (T3) under surface drip irrigation system. Water stress affected N, P and K uptakes. Reducing the soil water content decreased P uptake because it diminished the movement of P to roots by reducing the thickness of water films [2].

10.4 SUMMARY

Irrigation frequency is one of the most important factors in drip irrigation scheduling that affects the soil water regime, the water use efficiency, fertilization use efficiency, and the crop yield. Therefore, two field experiments were conducted in the summer season of 2011 and 2012 on sandy soil to investigate the effects of irrigation frequency and surface and subsurface drip irrigation systems on growth parameters, grain yield, N, P and K uptakes, and water use efficiency (WUE) of maize (*Zea mays* L.). The results indicated that the highest values of maize growth parameters were gained by irrigating plants with 100% of the ETc (2500 m³/fed. = control) treatment under subsurface drip irrigation system. On the contrary, the lowest values were observed by irrigating plants at 1900 m³/fed. (70% of ETc) under surface drip irrigation system. There were no significant differences among growth parameter values using the two experimental irrigation systems, but the differences were increased by using three water quantities. For example, dry weight of whole plant decreased by 46% and 35% under subsurface drip irrigation system, and by 36% and 27% under surface drip irrigation system comparing to other two water quantities (80% of ETc and 70% of ETc), respectively. The highest values of maize grain yield (2638 and 2575 kg/fed.) were gained using control irrigation water quantity (2500 m³/fed.) comparing to other two water quantities under subsurface and surface drip irrigation system, respectively. The relationship between water quantities and maize grain yield using drip irrigation system under semiarid conditions was linear.

KEYWORDS

- crop evapotranspiration, ETc
- deficit irrigation
- fertilization use efficiency
- grain yield
- growth parameter
- irrigation frequency
- irrigation scheduling
- irrigation system
- K uptake
- maize
- N uptake
- Nile Delta
- P uptake

- **regression analysis**
- **regression coefficient**
- **semiarid**
- **statistical model**
- **subsurface drip irrigation**
- **surface drip irrigation**
- **water scarcity**
- **water use efficiency, WUE**

REFERENCES

1. Achtnich, W. (1980). Bewa¨sserungslandbau: Agrotechnische Grundlagen der Bewasserungs-landwirtschaft. *Verlag Eugen Ulmer*, Stuttgart.
2. Ahmed, Amal O., Mekki, B. B. (2005). Yield and yield components of two maize hybrids as influenced by water deficit during different growth stages. *Egypt. J. Appl. Sci.*, 20:64–79.
3. Allen, R. G., Pereira, L. S., Raes, D., Smith, M. (1998). *Crop evapotranspiration: guidelines for computing crop water requirements*. FAO Irrigation and drainage paper 56. Rome, Italy: FAO.
4. Allen, R. G., Smith, M., Pruitt W. O., Pereira, L. S. (1996). Modifications to the FAO crop coefficient approach. In: *proceedings of the international conference on evapotranspiration and irrigation scheduling*, San Antonio, Texas, USA, pp. 124–132.
5. Assouline, S. (2002). The effects of micro drip and conventional drip irrigation on water distribution and uptake. *Soil Sci. Soc. Am. J.*, 66:1630–1636.
6. Batanouny, K. H., Hussein, M. M., Abo El Kheir, M. S. A. (1991). Response of *Zea mays* to temporal variation of irrigation and salinity under farm conditions in the Nile Delta, Egypt. *Proc. Inter. Conf. Plant Growth, Drought and Salinity in the Arab Region.* pages 189–204.
7. Blum, A., B. Arkin, 1984. Sorghum root growth and water use as affected by water supply and growth duration. *Field Crops Res.*, 9, 131–142.
8. Coelho, E. F., Or, D. (1999). Root distribution and water uptake patterns of maize under surface and subsurface drip irrigation. *Plant Soil*, 206, 123–136.
9. Davies, W. J., Zhan, J. (1991). Root signals and the regulation of growth and development of plants in drying soil. *Ann. Rev. Plant Physiol.*, 42, 55–76.
10. El-Noemani, A. A., Abd El-Halim, A. K., El-Zeiny, H. A. (1990). Response of maize (*Zea mays* L.) to irrigation intervals under different levels of nitrogen fertilization. *Egypt J. Agron.*, 15, 147–158.
11. El-Sheikh, M. A. (1994). Response of two maize varieties to plant densities and irrigation treatments. *Agric. Sci. Mansoura Univ., Egypt*, 19, 413–422.
12. Fahmy, S., Ezzat, M., Shalby, A., Kandil, H., Sharkawy, M., Allam, M., Assiouty, I., Tczap, A. (2002). *Water Policy Review and Integration Study*. Report No. 65 by Ministry of Water Resources and Irrigation, Egypt.
13. Farhad, W., Saleem, M. F., Cheema, M. A., Hammad, H. M. (2009). Effect of poultry manure levels on the productivity of spring maize (*Zea mays* L.). *J. Anim. Plant Sci.*, 19, 122–125.
14. Farooq, M., Wahid, A., Kobayashi, N., Fujita, D., Basra, S. M. A. (2009). Plant drought stress: effects, mechanisms and management. *Agron. Sustain. Dev.*, 29, 185–212
15. Goyal, Megh R. (2015). *Research Advances in Sustainable Micro Irrigation*. Volumes 1 to 10. Oakville, ON, Canada, Apple Academic Press Inc.,
16. Grant, R. F., Jakson, B. S., Kinuy, J. R., Arkin, F. (1989). Water deficit timing effects on yield components in maize. *Agron. J.*, 81, 61–65.

17. Howell, T. A., Yazar, A., Schneider, A. D., Dusek, D. A., Copeland, K. S. (1995). Yield and water use efficiency of maize in response to LEPA irrigation. *Trans. ASAE*, 38, 1737–1747.
18. ICID, (1998). Contemporary Challenges for Irrigation and Drainage. *Proceedings from the USCID 14th Technical Conference on Irrigation, Drainage and Flood Control.* Phoenix, Arizona, June 3–6, pp. 99–116. USCID, Denver, CO.
19. Jaleel, C. A., Manivannan, P., Wahid, A., Farooq, M., Al-Juburi, H. J., Somasundaram, R., Panneerselvam, R. (2009). Drought stress in plants, A review on morphological characteristics and pigments composition. *Int. J. Agric. Biol.,* 11, 100–105
20. Kanber, R., Yazar, A., O"nder, S., Ko"ksal, H. (1993). Irrigation response of Pistachio. *Irrig. Sci.,* 14, 1–14.
21. Kramer, P., Boyer, S. (1995). *Water Relations of Plants and Soils.* Academic Press, New York.
22. Mahrous, N. M. (1991). *Performance of some corn cultivars under some water stress treatments.* Bull. Fac. Agric., Cairo Univ., 42, 1117–1132.
23. Molden, D. J., El-Kady, M., Zhu, Z. (1998). Use and productivity of Egypt's Nile water. In, J. I. Burns, S. S. Anderson (eds.), *Proceedings from the USCID 14th Technical Conference on Irrigation, Drainage and Flood Control.* Phoenix, Arizona, June 3–6, pp. 99–116. USCID, Denver, CO.
24. Moon, D., Gulik, W. V. (1996). Irrigation scheduling using GIS. In, *Proceedings of the International Conference on Evapotranspiration and Irrigation Scheduling*, Nov. 3–6, pp. 644–649. Am. Soc. Of Agric. Eng., San Antonio, TX.
25. Neale, C. M. U., Ahmed, R. H., Moran, M. S., Pinter, J. P., Qi, J., Clarke, T. R. (1996). Estimating cotton seasonal evapotranspiration using canopy reflectance. In, *Proceedings of the International Conference on Evapotranspiration and Irrigation Scheduling*, Nov. 3–6, pp. 173–181. Am. Soc. Of Agric. Eng., San Antonio, TX
26. Phene, C. J., Davis, K. R., Mead, R. M., Yue, R., Wu, I. P., Hutmacher, R. B. (1994). Evaluation of a subsurface drip irrigation system after ten cropping seasons. ASAE Paper 93–2560 (winter meeting, Chicago, IL). ASAE, St. Joseph, MI.
27. Quaranta, F., E. Desideno and M. Magini, (1998). Summer sown maize comparison trial of 35 early hybrids. *Inform. Agron.*, 54, 55–59.
28. Rhoads, F. M., Bennett, J. M. (1990). Maize. In, B. A. Stewart, D. R. Nielsen, eds. *Irrigation of Agricultural Crops*, pp. 569–596. American Society of Agronomy, Madison, WI.
29. Simpson, G. M. (1981). *Water Stress on Plant.* Praeger Publishers CBS, Educational and Professional Publishing New York, USA.
30. Smith, J. A. C., Griffiths, H. (1993). *Water Deficits, Plant Responses from Cell to Community.* UK, Bios Scientific Publishers, pp. 331–332.
31. Snedecor, G. W., Cochran, W. G. (1980). *Statistical Methods.* 7th Ed., Iowa State Univ. Press, Iowa, USA.
32. Song–Feng, B., Dai Jun, Y., Zhang Lie, H., Yi Qing, (1998). Effect of water stress on maize pollen vigor and filament fertility. *Acta Agron. Simca*, 24, 368–373.
33. Stewart, J. D., Zine El-Abidine A., Bernier, P. Y. (1995). Stomatal and mesophyll limitations of photosynthesis in black spruce seedlings during multiple cycles of drought. *Tree Physiol.,* 15, 57–64.
34. Stone, P. J., Wilson, D. R., Reid, J. B., Gillespie, G. N. (2001). Water deficit effects on sweet maize, I. Water use, radiation use efficiency, growth, and yield. *Aust. J. Agric. Res.*, 52, 103–113.
35. Turner, N. C., Toole, J. C. O., Cruz, R. T., Yambao, E. B., Ahmad, S., Narnuco, O. S., Dingkhum, M. (1986). Response of seven diverse rice cultivars to water deficits, Osmotic adjustment, leaf elasticity, leaf expansion, leaf death, stomatal conductance and photosynthesis. *Field Crops Res.,* 13, 273–286.
36. Wang, F. X., Kang, Y., Liu, S. P. (2006). Effects of drip irrigation frequency on soil wetting pattern and potato growth in North China Plain. *Agric. Water Manage.,* 79, 248–264.

CHAPTER 11

WATER DISTRIBUTION UNDER SURFACE EMITTER AND SUBSURFACE TEXTILE IRRIGATION

A. A. MADY, H. M. MEHANNA, and S. OKASHA

CONTENTS

Modified from A. A. Mady, H. M. Mehanna,, and S. Okasha, "Maximizing of Water Conservation in Sandy Soil Using Geotextile Material under Surface Drip Irrigation System." *Journal of Applied Sciences Research,* 8(8), 4014–4022, 2012. http://www.aensiweb.com/old/jasr/jasr/2012/4014-4022.pdf. Modified by the authors. Originally published via open access.

11.1   INTRODUCTION

Egypt is located in the semiarid region of Middle East, and is characterized by evaporation rate of 3000 mm/year and precipitation rate of 15.3 mm/year [2]. Therefore, Water resources are a limiting factor for agricultural production. Hence, water saving becomes clearly a necessary prerequisite. In this logic, irrigation technologies appear to play a major role for supporting agro-economy of Egypt.

El-Gindy [5] found that drip irrigation method increased the green pepper yield by 64% over traditional furrow irrigation method beside its higher water use efficiency (WUE). Shrestha and Gopalakrishnan [12] reported that drip irrigation was a preferred alternative, resulting in 15% increase in sugarcane yield and 12% reduction in water use efficiency. This superior performance of drip irrigation system was largely due to land improvement characteristics of drip irrigation technology. Hapase et al. [6] demonstrated that drip irrigation increased sugarcane yield by 37%, and a higher water use efficiency of 2.7 times compared to furrow irrigation. Anthony and Namoi [1] reported that the average yield advantage of surface drip irrigation was 8.5 kg lint cotton/ha compared to an additional yield of 60 kg/ha in buried drip irrigation.

Subsurface textile irrigation (SSTI) is a technology designed specifically for subsurface irrigation in all soil textures from desert sands to heavy clays. Use of SSTI will significantly reduce the usage of water, fertilizer and herbicide. It will lower on-going operational costs and, if maintained properly, will last for decades. By delivering water and nutrients directly to the root zone, plants are healthier and have a far greater yield [7]. It is the only irrigation system that can safely use recycled water or treated water without expensive "polishing" treatment because water never reaches the surface. A typical subsurface textile irrigation system has an impermeable base layer (usually polyethylene or polypropylene or polyester sheet), a drip line running along that base, a layer of geotextile on top of the drip line and, finally, a narrow impermeable layer on top of the geotextile (see diagram). Unlike standard drip irrigation, the spacing of emitters in the drip pipe is not critical as the geotextile moves the water along the fabric up to 2 m from the dripper. SSTI is installed 15–20 cm below the surface for residential/commercial applications and 30–50 cm for agricultural applications.

The geosynthetic material product is manufactured by Dalian Hengda high technology material development company in Egypt. It is a new type of building material applied to geotechnical engineering and civil engineering. Such Geotextile is constructed from polyester filament netting that is consolidated with fiber array of three dimensional structure under subsurface, thus imparting important property of leakage-proof material under the subsurface of the soil to prevent or reduce the leakage by deep percolation under different irrigation systems. Therefore, the polyester filament spun bonded needle-punched geotextiles are mainly used in national projects in Egypt (including expressway, railway, water conservancy, etc.) as well as water proof materials. Geotextiles also have many functions such as protection, partition, filtration, reinforcement, antiseepage and drainage functions, hence possessing high popularization. El-Gamal and El Shafey [4] found that drainage, filtration and reinforcement are the principle functions of geotextile products. For the drainage application, filtration occurs when a liquid passes through the plane of a geotextile while retaining soil particles on the upstream side of the fabric, providing a soil filtration system similar to traditional graded aggregate structure. One of the uses of the geotextile material is in the geotube technology that can reduce total suspended solids concentrations, allowing filtrated water to travel through a T-Tape drip irrigation system without clogging the emitters. Jason [8] indicated that the geotube retained solids with high NPK value than either the pond sludge or geotube filtrate. The analysis of the retained solids showed that it was composed of 59.3% organic matter with 86.4% moisture content after four weeks of dewatering). Therefore, geotextile material has another propriety that it can retain a percentage of NPK, which goes down with deep percolation.

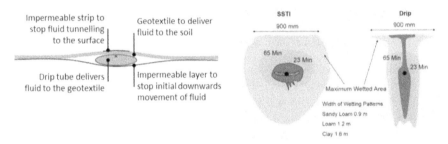

Figure 11.1   Typical example of SSTI installation (left), with a typical pattern under SSTI and drip irrigation system alone (right), [7]

Najafi and Tabatabaei [11] illustrated that application of sand and geotextile envelope in subsurface drip irrigation emitter is a new suggestion for controlling the clogging of emitters. Sand envelope around the emitter has been recommended for deeper root zone and long-term use. Geotextile envelope is suitable for seasonal crops with shallow rooting system. In addition, the above-mentioned filters can provide better hydraulic conductivity around the emitter. Lanjabi – Sharahi [9] detected that contact between soil and the geotextile envelope may push soil moisture pattern to distribute more horizontally in a sandy loam soil and high-density geotextile.

This chapter discusses the effects of using the geotextile material as a semipermeable layer under the emitters in surface drip irrigation on soil moisture distribution in sandy soil.

11.2   MATERIALS AND METHODS

At the experimental site (El-Nagah village, El-Bustan region, El-Beheira Governorate), soil texture was sandy soil. The top vegetative cover of the soil was scraped away and the soil from the top 0–50 cm was removed with an auger. Soil samples were taken from 3 cores with an auger for bulk density determination. Five samples were taken (every 10 cm soil depth up to 50 cm). The soil was placed in heavy duty polyethylene bags for each layer. The bags were transported from the collection site to the laboratory. Table 11.1 presents some physical properties of soil layers. Soil samples were analyzed according to Black et al. [3].

TABLE 11.1 Some Physical Properties of Soil at Different Depths

Depth cm	Grav- el, %	Sand, %	Silt, %	Clay, %	Bulk density, g/cm3	Poros- ity, %	F.C., %	Soil texture
0 – 25	3.61	78.44	10.31	7.64	1.40	46.15	14.00	Sandy
25–37.5	0.80	86.12	8.29	4.79	1.52	41.54	13.31	Sandy
37.5–50	0.65	89.22	6.07	4.06	1.56	40.00	12.05	Sandy

11.2.1 THE LABORATORY APPARATUS

Experiments were conducted in a wooden rectangle soil box (50 length, 27 width, and 70 cm depth), which was painted with water proof paint to avoid water absorption. The front side of this box was transparent glass to allow visual observations of the water movement along the wetting front [13]. In this laboratory experiment, 3 boxes were used, with soil layers in the same order as in the site environment. The dripper line was put on the soil surface in the middle of the experimental box. The specifications of the geotextile material are presented in Table 11.2. This material was placed at two depths (25 and 37.5 cm), in the subsurface of the soil which

TABLE 11.2 The Technical Specifications of the Geotextile Sheet

Technical properties	Value
Width, m	4
Mass g/m2	150 ± 10%
Thickness, mm	1.7
Tensile strength, kN/m	7.5
Tensile elongation, %	30 – 80

11.2.2 EXPERIMENTAL PROCEDURE

The soil was air-dried. The large clods were removed by passing the soil through 2 mm mesh size sieve. The soil was placed into the soil box in 5 cm layers to achieve the same bulk densities as measured in the field and leveled until the target depth (25 cm and 37.5 cm from the box surface) to add the thin layer of the geotextile material, which was placed at 25 or 37.5 cm depth [10]. The water was supplied from a water tank by using pump connected to it with 0.008 m³/hr. flow rate through GR dripper line. The wetting front contour was observed through the transparent glass wall and projected on the transparent paper sheet. Soil samples were taken at 5 cm apart along the depth of the soil layers (24 and 48 h after each irrigation) in the vertical direction parallel to the transparent glass wall to represent the moisture distribution. The dripper line was exactly located at the center at the soil surface. Surfer program Version 7 (Surface Mapping System, 1999) was used to draw both of water and salt distributions in the soil profile.

11.2.3   FIELD EXPERIMENT

The field experiment was located at. The farm was planted with Naval orange one year before the initiation of research for water and salt movement. Row spacing was 5 m and plant-to-plant spacing was 4 m. Irrigation depth was calculated based on the climatic data for this location taking an average of the five year data. Crop data and field data was recorded during the 2012 growing season.

11.3   RESULTS AND DISCUSSION

11.3.1   WITHOUT GEOTEXTILE SHEET

The effects of soil texture on the moisture distribution were recorded once 24 and 48 h after the start of first/ second/ third irrigation under surface drip irrigation line as shown in Figs. 11.2 to 11.4 for the first, second and third irrigations without using the geotextile material sheets, respectively.

It is concluded from Fig. 11.2 that the contour lines were closely spaced to each other at 24 h after first irrigation (one hour operating time or 8 L of water). Also, the maximum moisture content was directly under the dripper. Figure 11.2a shows that the moisture content decreased by increasing the soil depth and by increasing the horizontal distance from the dripper. It was observed that the moisture distribution was very symmetrical. Figure 11.2b shows that the moisture contour lines moved downward 48 h after the irrigation. Therefore, the moisture contour lines were widely spaced from each other more compared to those 24 h after the irrigation. This action of water movement was expected, but the moisture content decreased in the first 10 cm of the soil surface.

Figure 3 shows that the moisture distribution under the dripper 24 and 48 h after the start of second irrigation, respectively. The moisture content was concentrated under the dripper. There was excessive water, which was lost by deep percolation (Fig. 11.3a). Nearly 50 cm from the soil surface, the maximum moisture content was observed. On the contrary, Fig. 11.3b shows a different trend of water movement, where the water moved toward the soil surface following the capillary rise theory, especially under the dripper, and the moisture distribution was observed more than after one day after the start of second irrigation. It was also observed that moisture exited in the soil layers before the second irrigation (approximately the soil was semisaturated).

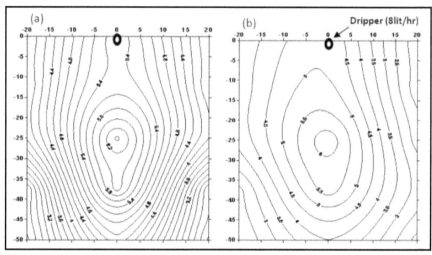

FIGURE 11.2 Soil moisture distribution contour lines at different depths of the soil box 24 h after the start of first irrigation (Fig. 11.2a, left) and 48 h after the start of first irrigation (Fig. 11.2b, right): Without geotextile sheets.

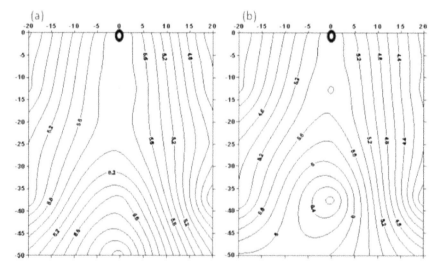

FIGURE 11.3 Soil moisture distribution contour lines at different depths of the soil box 24 h after the start of second irrigation (Fig. 11.3a, left) and 48 h after the start of first irrigation (Fig. 11.3b, right): Without geotextile sheets.

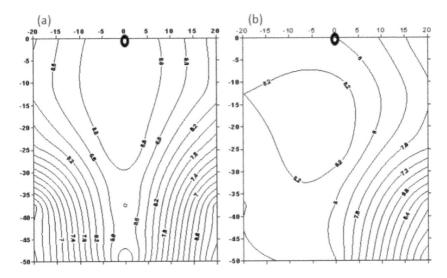

FIGURE 11.4 Soil moisture distribution contour lines at different depths of the soil box 24 h after the start of third irrigation (Fig. 11.4a, left) and 48 h after the start of third irrigation (Fig. 11.4b, right): Without geotextile sheets.

Moisture contour lines for third irrigation are shown in Fig. 11.4. Figure 11.4a indicates that there was excess of water content in the soil layers, the moisture content was increased symmetrically under the dripper more than that occurred after the second irrigation or the first irrigation. Near 50 cm depth from the soil surface, existence of deep percolation was observed. On the contrary, Figure 11.4b shows an irregular shape or distribution of moisture in the soil profile 48 h after the start of irrigation. Generally, it is noted that there was a water movement 48 h after irrigation toward the left side from the dripper or the center of box and downward in the soil profile. From Figs 11.2–4, it can be concluded that the soil held moisture from one irrigation to another; and by increasing the irrigation water, the water moved downward to deep percolation. Therefore, moisture content decreased in the upper soil layer especially 48 h after the start of irrigation.

11.3.2 THE EFFECT OF ADDING GEOTEXTILE SHEET AT 25 CM SOIL DEPTH ON THE MOISTURE DISTRIBUTION

Figures 11.5–11.7 show the contour lines for moisture distribution 24 and 48 h after initiation of first, second, third irrigation under surface drip irrigation, with geotextile sheet at 25 cm soil depth.

Figures 11.5a and 11.5b show the moisture distribution 24 and 48 h after the start of first irrigation, respectively. The moisture contour lines were symmetrical under the dripper and were widely spaced in the upper soil over the geotextile sheet and were closely spaced below the geotextile sheet in the soil profile. The lowest moisture content was observed at 25 cm under the soil surface. On the other hand, the highest moisture content was observed at 20 – 25 cm from the soil surface directly above the geotextile sheet. This occurred at 24 and 48 h after the addition of water by the dripper. Generally, the moisture contour lines were distributed symmetrically. The moisture was decreased clearly below the geotextile sheet. Therefore, geotextile sheet minimized deep percolation, which occurs mainly after the irrigations.

Figure 11.6 shows the moisture distribution for the second irrigation. In Figs. 11.2–11.5, it was observed that the moisture content was increased from one irrigation to another, because the soil moisture was present from the previous irrigation. The moisture contour lines were closely spaced to each other compared to those obtained in the soil profile without the geotextile sheet (Figs. 3a and 4a) at 24 h after the irrigation.

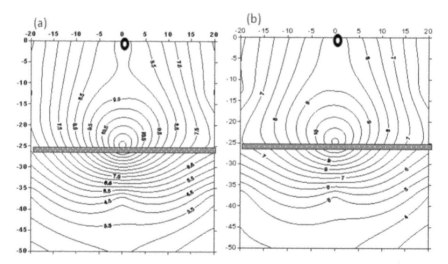

FIGURE 11.5   Moisture distribution contour lines at different soil depths at 24 h (Fig. 11.5a, left) and 48 h (Fig. 11.5b, right) after the start of first irrigation with geotextile sheet at 25 cm soil depth.

The moisture contour lines in Figs. (6-b) and (6-b) were a little bit far from each other at 48 h after the second irrigation as well as the third irrigation, respectively. It was observed that the moisture content decreased by time without irrigation, and the moisture content above the Geotextile sheet (at 25 cm depth into the soil box) was greater than that indicated under it for the two mentioned irrigations.

From Figs. (7-a) and (7-b), it is worth mentioning that the moisture contour lines were regular by using the Geotextile sheet at 25 cm depth in the soil profile rather than that obtained without the Geotextile material in the soil profile.

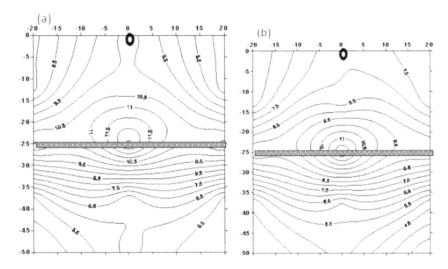

FIGURE 11.6   Moisture distribution contour lines at different soil depths at 24 h (Fig. 11.6a, left) and 48 h (Fig. 11.6b, right) after the start of second irrigation with geotextile sheet at 25 cm soil depth.

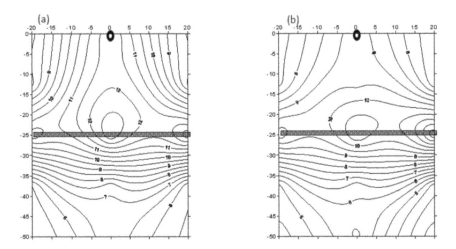

FIGURE 11.7   Moisture distribution contour lines at different soil depths at 24 h (Fig. 11.7a, left) and 48 h (Fig. 11.7b, right) after the start of third irrigation with geotextile sheet at 25 cm soil depth.

11.3.3 THE EFFECTS OF ADDING GEOTEXTILE SHEET AT 37.5 CM SOIL DEPTH ON MOISTURE DISTRIBUTION BELOW THE SURFACE EMITTER

Figures 11.8a–11.10a show the soil moisture contour lines under the surface drip-per a fixed geotextile sheet at 37.5 cm from the soil surface, at 24 h after the start of first, second and third irrigations, respectively. Irregular shapes of the moisture contour lines can be observed in the first and the second irrigations, however at 24 after the start of third irrigation, there were semisymmetrical moisture contour lines. The differences were small between the moisture content above and under the geotextile sheet (at 37.5 cm depth from the soil surface). There was a horizontal movement of the irrigation water in the soil profile. On the contrary, Figures 11.8b–11.10b illustrate the semisymmetrical moisture contour lines at 48 h after the 1st, 2nd and the 3rd irrigations, respectively. The difference was pronounced between the soil moisture content above and below the geotextile sheet at 48 h after the start of third irrigation (Fig. 10b). Generally, installing the geotextile sheet at 37.5 cm depth requires changes in the amount of water applied or increasing the time of irrigation in order to get the most benefit from the geotextile sheet.

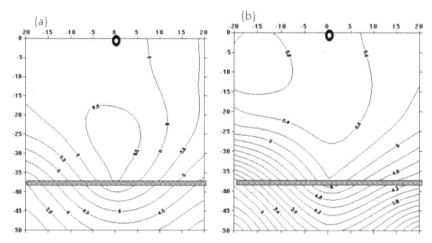

FIGURE 11.8   Moisture distribution contour lines at different soil depths at 24 h (Fig. 11.8a, left) and 48 h (Fig. 11.8b, right) after the start of first irrigation with geotextile sheet at 37.5 cm soil depth.

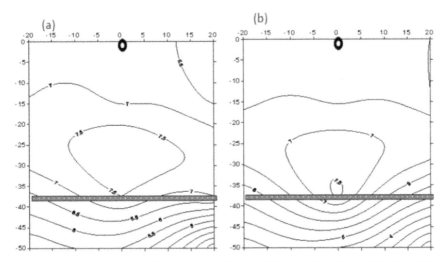

FIGURE 11.9 Moisture distribution contour lines at different soil depths at 24 h (Fig. 11.9a, left) and 48 h (Fig. 11.9b, right) after the start of second irrigation with geotextile sheet at 37.5 cm soil depth.

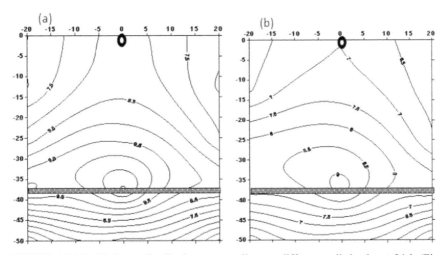

FIGURE 11.10 Moisture distribution contour lines at different soil depths at 24 h (Fig. 11.10a, left) and 48 h (Fig. 11.10b, right) after the start of third irrigation with geotextile sheet at 37.5 cm soil depth.

11.3.4   FIELD EXPERIMENT IN DRIP IRRIGATED NAVAL ORANGE

The results obtained from the laboratory experimental in Section 11.3.1–11.3.3 in this chapter indicated that using of geotextile sheets at 25 cm depth under the soil surface was more effective on the moisture distribution for the different irrigations than using at 37.5 cm depth or deeper. Therefore, in the field experiment, geotextile sheets were buried at 25 cm depth in the experimental trees. The moisture distribution contour lines were drawn for three irrigations in the productivity period of the Navel orange trees. The amounts of water in three irrigations were equaled for the experimental treatments (without Geotextile sheets (control) and fixed Geotextile sheets at 25 cm depth under the trees).

Figures 11.11–11.13 show the moisture distribution contour lines at different levels of the soil profile at 24 h after the start of 1st, 2nd, and 3rd irrigations, respectively. Generally, using the geotextile sheets at 25 cm depth improved the moisture distribution in the sandy soil by decreasing the deep percolation. The moisture was distributed symmetrically. Therefore, the irrigation water was conserved with the use of geotextile sheets, avoiding water loss by deep percolation. Figure 11.14 shows the effect of using geotextile sheets at 25 cm depth on the number of fall fruits per tree and the yield of Naval orange (kg/ tree). Data illustrated that using geotextile sheets at 25 cm depth decreased the number of fruits per tree by 51.22%, and increased the fruit yield by 23.54% comparing to the control treatment (no geotextile sheet).

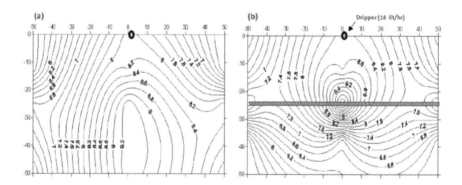

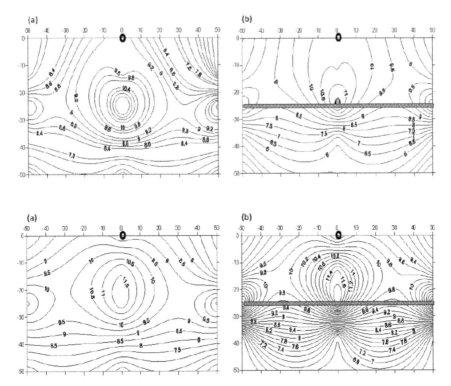

FIGURES 11.11–11.13 Moisture distribution contour lines at different soil depths at 24 h after start of irrigation. Fig. a: Without geotextile sheet; Fig. b: a fixed Geotextile sheet at 25 cm depth from the soil surface. Fig. 11.11. First irrigation (top); Fig. 11.12. Second irrigation (center); Fig. 11.13. Third irrigation (bottom).

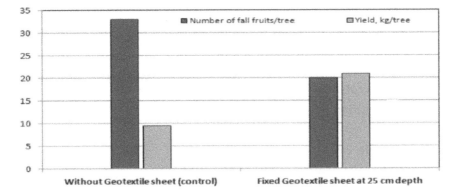

FIGURE 11.14 The effects of using geotextile sheets at 25 soil depth on fall number of fruits/tree and fruit yield/tree, under drip irrigated Naval orange, during the fall.

11.4   CONCLUSIONS

Using geotextile sheets at 25 cm soil depth was more effective than the sheets at 37.5 cm or deeper; caused reduction in losses due to deep percolation. The moisture content in 0–25 cm soil depth was higher for geotextile sheet at 25 cm depth than at deeper depths. Generally, geotextile sheets at 37.5 cm depth from the soil surface needed a considered amount of irrigation water so that moisture distribution was more regular and the use of sheets was more efficient. Using geotextile sheets at depths greater than 25 cm is not effective to save water. Therefore, it is recommended to install geotextile sheets at 25 cm depth to save water and to allow horizontal water movement in the upper layer of the soil needed for plants growth. Using geotextile sheets at 25 cm depth under navel orange trees improved the moisture distribution in the sandy soil by decreasing the deep percolation; decreased number of fruits per tree by 51.22%; and increased fruit yield of Naval orange by 23.54% comparing to the control treatment (no geotextile sheets).

11.5   SUMMARY

A laboratory experiment was designed to study the effects of geotextile sheets at two soil depths on water distribution in the soil profile, under a surface emitter. Geotextile is made of polyester filament netting and by consolidating with fiber array of three-dimensional structure. It has clear mechanical properties, excellent longitudinal and transverse drainage properties as well as excellent resistance against aging, acids, alkalis and biological attacks. Results indicated that geotextile sheets at 25 cm depth were more effective than using at 37.5 cm or deeper. In such case, the deep percolation can be reduced, and the moisture content was higher above the geotextile sheet at 25 cm depth than under it. Using of the geotextile sheet deeper than 25 cm was not effective to save water. Therefore, using sheets at 25 cm is recommended for water saving and greater horizontal water movement in the upper layer of the soil which is needed for plant growth.

KEYWORDS

- deep percolation
- drip irrigation
- emitter
- geotextile
- moisture distribution
- sandy soil

- **soil depth**
- **soil texture**
- **subsurface textile irrigation**
- **surface drip irrigation**
- **textile**
- **water distribution**

REFERENCES

1. Anthony, D., Namoi, A. (1996). Drip irrigation: separating fact from fiction. *Australian Cotton Grower*, 17(3), 60–66.
2. AOAD, (2002). *Report of maximizing using water harvesting techniques*. Arab Organization of Agricultural Development (AOAD). Pages23.
3. Black, C. A., Evans, D. P., Ensminger, L. E., White, J. L., Clark, F. E. (1982). *Methods of soil analysis*. Madison, WI: American Society Agronomy.
4. El-Gamal, M. A. L., El-Shafey, M. (2000). Performance of drainage geotextiles for sustainable development of soil and water resources in Egypt. In: *El-Gamal* et al., *ICEHM* (2000). Cairo University, Egypt, September, pages 1- 13.
5. El-Gindy, A. M. (1984). Optimization of water use efficiency for pepper crop. *Annals of Agric. Sci., Fac. of Agric.,* Ain Shams Univ., Cairo, 29(1), 539–555.
6. Hapase, D., Mankar, A., Salunkle, A. (1992). Techno-economic of drip irrigation for sugarcane. *Crop. Inter. Agric. Conf.*, AIT, Bangkok, (33), 897–904.
7. http://en.wikipedia.org/wiki/Subsurface_textile_irrigation
8. Jason, D. (2008). Evaluating geotextile technology to enhance sustainability of agricultural production systems in the U. S. Virgin Islands. *Aquaponics Journal*, www.aquaponicsjournal.com, 3rd Quarter.
9. Lanjabi-Sharahi, M. (2009). Soil wetting pattern under subsurface drip irrigation with a geotextile envelope. MSc Thesis, Department of Water Engineering, Faculty of Agriculture, Shahrekord University, Shahrekord, Iran.
10. Malik, M., Amrhein, C., Letey, J. (1991). Polyacrylamide to improve water flow and salt removal in a high shrink-swell soil. *Soil Sci. Am. J.,* 55:1664–1667.
11. Najafi, P., Tabatabaei, S. H. (2010). Application of sand and geotextile envelope in subsurface drip irrigation. *African Journal of Biotechnology*, 9(32), 5147–5150.
12. Shrestha, R. B., Gopala Krishnan, C. (1991). Water use efficiency under two irrigation technologies-a quadratic production function analysis. *International Journal of Water Resources Development*, 7(3), 133–137.
13. Zin El-Abedin, T. K. (2006). Improving moisture distribution pattern of subsurface drip irrigation in sandy soil by using synthetic soil conductors. *Misr J. Ag. Eng.,* 23(2), 374–399.

CHAPTER 12

SALT DISTRIBUTION UNDER SUBSURFACE TEXTILE IRRIGATION AND DRIP IRRIGATED CITRUS

H. M. MEHANNA, E. M. OKASHA, and M. A. A. ABDOU

CONTENTS

Modified from H. M. Mehanna, E. M. Okasha, and M. A. A. Abdou. "Optimization of Fertigation Process under Drip Irrigation System by Using Geotextile Sheets in Sandy Soil." *World Applied Sciences Journal*, 27(6), 688–693, 8(8), 2013. http://www.idosi.org/wasj/wasj27(6)13/1.pdf. Modified by the authors. Originally published via open access.

12.1   INTRODUCTION

Drip or trickle irrigation is a very efficient method of applying water and nutrients to crops. Crop yields can increase through improved water and fertilization management, and reduced disease and weed growth. When drip irrigation is used with polyethylene mulch, yields can increase even further. The ability to "spoon-feed" nutrients along with water is partially responsible for the yield increases resulting from drip irrigation. Three types of chemicals are typically injected into drip irrigation systems: fertilizers, pesticides and anticlogging agents. Fertilizers are the most common. Positive displacement, pressure differential and pump powered injectors are common injectors for chemigation [4]. The subsurface drip irrigation (SDI) system is flexible and can provide frequent light irrigations. This is especially suitable for arid, semiarid, hot and windy regions with limited water supply. Farm operations also become free of impediments that normally exist above ground with any other pressurized irrigation system [8]. Roots and clay particles can clog drip emitters in systems buried below mulch or below the soil surface and this tendency historically limits the service life of the system [8]. Tabatabaei and Najafi [9] reported that prevalence contamination (such as $N-NO_3$) is minimized due to usage of treated wastewater (TWW) in depth and groundwater, in the case of SDI due to minimal leaching [2].

Sandy soils are found in all parts of the world [1, 10]. When these soils have greater than 70% sand and less than 15% clay, they are classified as Arenosols in the World Reference Base system [1, 10]. Characteristic properties of Arenosols are high water permeability, low water-holding capacity, low specific heat, and often-minimal nutrient contents. Arenosols have a large number of unfavorable attributes for sustainable agriculture [6, 7].

Lanjabi Sharahi [5] detected that geotextile envelope may push soil moisture patterns to distribute more horizontally in a sandy loam soil. Mady et al. [6] illustrated that use of geotextile sheets at 25 cm soil depth was more effective than using at 37.5 cm or deeper. Use of geotextile sheets deeper than 25 cm was not effective to save water. Jason [3] indicated that the geotube-retained solids contained a higher NPK value than either the pond sludge or geotube filtrate. Therefore, Geotextile material has another characteristic of retaining of NPK nutrients.

The research study aimed to recommend solutions to problems of washing and leaching of fertilizers by deep percolation in sandy soils. Authors evaluated the effects of installation of the geotextile materials in the shape of sheets at different soil depths to optimize the installation depths. They also studied effects of these sheets on the distribution of fertilizers in the soil profile. The reader may also like read Chapter 11 on moisture distribution under an emitter with geotextile sheets.

12.2 MATERIALS AND METHODS

The experiment was conducted at a private farm in El-Nagah village, El-Bustan region, El-Beheira Governorate. Soil texture is sandy. Table 12.1, of Section 11.2 in Chapter 11 in this book, presents some physical properties of the soil layers by analyzing soil samples according to Page et al. [7]. Table 12.1 shows the chemical analysis of soil layers and the irrigation water at the site. The specifications of the geotextile material are presented in Table 12.2. This material was placed at two soil depths (25 and 37.5 cm).

TABLE 12.1 Chemical Analysis of Soil Site and Irrigation Water (Analyzed by Soil and Water Laboratory)

Item	Depth, cm	pH (1:2.5)	EC, ds/m (1:5)	Soluble cations, mq/L				Soluble anions, mq/L			
				Ca^{++}	Mg^{++}	Na^{++}	K^+	CO_3	HCO_3	Cl^-	SO_4
Soil	0–25	7.70	0.41	1.22	0.7	0.54	0.51	0.64	0.91	0.8	0.62
	25–37.5	7.60	0.32	1.24	0.5	0.41	0.47	0.5	0.90	0.7	0.52
	37.5–50	7.59	0.22	0.9	0.5	0.46	0.36	0.21	1.00	0.8	0.21
Water	—	7.54	0.34	1.43	0.57	2.45	0.19	0	1.11	2.79	0.74

TABLE 12.2 The Specifications of the Geotextile Material

Technical properties	Value
Width, m	4
Mass, g/m^2	150±10%
Thickness, mm	1.7
Tensile strength, kN/m	7.5
Tensile elongation, %	30–80

12.2.1 THE EXPERIMENTAL PROCEDURE

Soil samples were taken at intervals of 10 cm soil depth up to 50 cm; and at every 10 cm in the horizontal direction on two sides of the dripper. The soil samples were air dried on the dry weight. The electrical conductivity (EC, as an indicator of the fertilizers distribution in the soil profile) of soil samples extract was used to find the percentage of soluble salts. The samples were saturated with distilled water, and were vacuum filtered to separate water from the soil. Then EC of the saturated paste extract was measured with EC meter. Soil samples were taken at 24 h before and after two fertigations. Surfer program Version 7 (surface mapping system, 1999) was used to draw both of water and salt distributions in the soil profile.

12.3   RESULTS AND DISCUSSION

12.3.1   USE OF GEOTEXTILE SHEETS BELOW AN EMITTER: SECOND FERTIGATION CYCLE

Figure 12.1 indicates contour lines for the distribution of salinity (EC, as an indicator of the fertigation process and the availability of fertilizers to the plant) in the soil profile (100 cm width in the horizontal direction, and 50 cm depth in the vertical direction) without geotextile sheets under the drip irrigated Navel orange trees. Figures 12.2 and 12.3 shows the contour maps for EC with geotextile sheets at 25 cm and 37.5 cm soil depths, respectively. Figures 12.2a and 12.2b illustrates the salinity distribution contour lines 24 h before and after the 2nd fertigation, respectively. After comparing data in Figs. 12.1–12.3, the salinity (EC) was increased after the fertigation cycle.

Data in Fig. 12.1 also indicates that the EC value was decreased more in the 0–25 cm soil layer than the values in next layers. This may be due to washing of nutrients by the irrigation water and then down with water to deeper layers by deep percolation. The salinity concentration increased in 25–50 cm depth soil layer by 30% more than in the 0–25 cm depth soil layer (top soil layer).

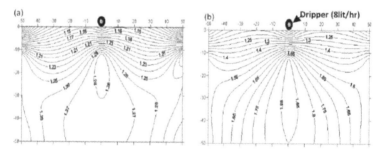

FIGURE 12.1   Salinity distribution (EC) contour lines in the soil profile 24 h before (Fig. 12.1a) and after (Fig. 12.1b) the second fertigation cycle, with no geotextile sheets.

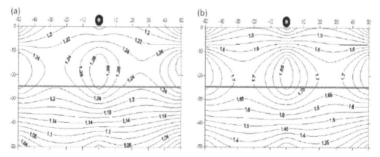

FIGURE 12.2   Salinity distribution (EC) contour lines in the soil profile 24 h before (Fig. 12.2a) and after (Fig. 12.2b) the second fertigation cycle, with geotextile sheets at 25 cm soil depth.

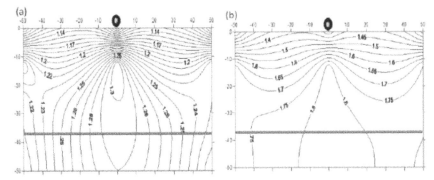

FIGURE 12.3   Salinity distribution (EC) contour lines in the soil profile 24 h before (Fig. 12.1a) and after (Fig. 12.1b) the second fertigation cycle, with geotextile sheets at 37.5 cm soil depth.

Figure 12.2 shows salinity distribution (EC) contour lines in the soil profile 24 h before (Fig. 12.2a) and after (Fig. 12.2b) the second fertigation cycle, with geotextile sheets at 25 cm soil depth. By observing the salinity distribution contour lines in Fig. 12.2b, we can conclude that the contour lines are symmetrically distributed in the horizontal direction. These results are in agreement with those by Mady et al. [6] who reported similar findings while studying the moisture distribution under surface drip irrigation system.

The values of EC in Fig. 12.3 indicates that the salinity increased after the second fertigation cycle and also the concentration was high in the soil layer (0–25 cm depth) above the geotextile sheets (fixed at 37.5 cm depth under the soil surface) than in the soil layer 25–50 cm below geotextile sheets (Fig. 12.3b). The highest values of salinity concentration were in 0–37.5 cm soil layer. Generally by comparing the salinity contours in the Figs. 1b to 3b, use of geotextile sheets at 25 cm depth increased the salinity concentration in the top soil layer (0–25 cm) more than that obtained without using these sheets or using them at 37.5 cm depth. Moreover, the differences were significant between the EC values between these sheets at 25 depth and 37.5 cm depth.

From Figs. 12.1–12.3, it can be concluded that the geotextile sheets at 25 cm depth were more effective for vegetable crops (shallow roots) compared to sheets at >37.5 cm depth for tree crops to save water and fertilizers in the active root zone [6].

12.3.2   USE OF GEOTEXTILE SHEETS BELOW AN EMITTER: THIRD FERTIGATION

In these treatments, geotextile sheets were fixed at 25 and 37.5 cm soil depths under an emitter. The salinity concentrations were recorded at 24 h before and after the second fertigations for both depths.

Figure 12.4 shows the data in case of no geotextile sheets. Data in Fig. 12.4b indicates that the distribution of salinity was not symmetrical under the dripper in the root zone, and the salinity concentration increased at deeper soil layers far from the active root zone, due to washing of nutrients by deep percolation. On the other hand, data in Figs. 12.5b and 12.6b indicated that the salinity contour lines were distributed symmetrically in the active root zone of the trees on either side of the dripper (horizontally from 0 location). The salinity concentration increased above the geotextile sheet than below it, especially by fixing the sheets at 25 cm depth. Using these sheets at 37.5 cm depth was more appropriate to the root growth of orange trees in this age (one year), to allow tree roots to be distributed to >25 cm layers.

These results were confirmed by measuring the number of fall fruits per tree and tree productivity (kg/tree), as shown in Fig. 12.7, where the highest value of the number of fall fruits per tree and the lowest value of tree productivity were obtained without the geotextile sheets in the soil profile, while the sheets saved fertilizers in the active root zone at 24 h after fertigation cycle, thus preventing use of excessive fertilizers. From Fig. 12.7, it can be concluded that the geotextile sheets at 37.5 cm depth improved the yield per tree by 150% and 19% compared to that obtained without using these sheets or using it at 25 cm depth, respectively.

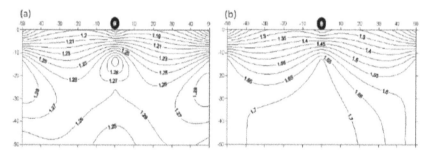

FIGURE 12.4   Salinity distribution (EC) contour lines in the soil profile 24 h before (Fig. 12.4a) and after (Fig. 12.4b) the third fertigation cycle, without geotextile sheets.

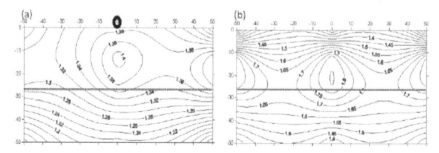

FIGURE 12.5   Salinity distribution (EC) contour lines in the soil profile 24 h before (Fig. 12.5a) and after (Fig. 12.5b) the third fertigation cycle, with the geotextile sheets at 25 cm depth.

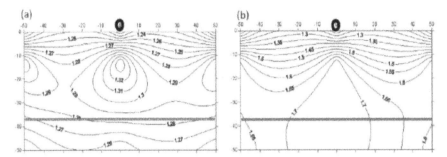

FIGURE 12.6   Salinity distribution (EC) contour lines in the soil profile 24 h before (Fig. 12.6a) and after (Fig. 12.6b) the third fertigation cycle, with the geotextile sheets at 37.5 cm depth.

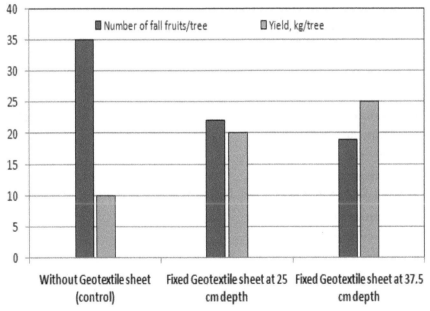

FIGURE 12.7   Effects of installation of geotextile sheets at 25 and 37.5 cm soil depths on the number of fall fruits/tree and fruit yield (kg/tree), for drip irrigated Naval orange orchard.

12.4   SUMMARY

This study was designed to evaluate the effects of placing geotextile material sheets at different depths on fertilizers distribution (using the electrical conductivity (EC) as an indicator) in the soil profile and on the performance of Navel orange trees (one year old) under surface drip irrigation system (SDI) in a sandy soil. Using of geotextile sheets at 25 cm depth increased the salinity concentration in the top soil layer (0–25 cm) more than that obtained without using these sheets or using these at 37.5 cm depth. Moreover there were no significant differences between using these sheets at 25 cm depth and 37.5 cm depth. Therefore, the geotextile sheets at 25 cm depth are more effective for vegetable crops (shallow roots), compared to sheets at >37.5 cm depths for tree crops to save water and fertilizers in the active root zone. Also results concluded that use of geotextile sheets at 37.5 cm and 25 depths improved the orange fruit yield per tree by 150% and 19%, respectively comparing to that obtained without using these sheets.

KEYWORDS

- deep percolation
- drip irrigation
- electrical conductivity, EC
- emitter
- fertigation
- geotextile
- moisture distribution
- salinity distribution
- sandy soil
- soil depth
- soil texture
- subsurface textile irrigation
- surface drip irrigation
- textile
- water distribution

REFERENCES

1. FAO, (2001). Lecture notes on the major soils of the world. World Soil Resources Reports 94. FAO, Rome.
2. Geoflow, (2009). Wastewater nano-rootguard. Available on internet, (www.Geoflow.com). Inc.
3. Jason Danaher, (2008). Evaluating geotextile technology to enhance sustainability of agricultural production systems in the U.S. virgin islands. Aquaponics Journal, www.aquaponicsjournal.com, 3rd Quarter, 2008

4. Lamont, W. J., Orzolek, M. D., Harper, K. J., Jarrett, A. R., Greaser, G. L. (2002). Drip irrigation for vegetable production. information and communication technologies in the college of agricultural sciences, Pennsylvania State University, (agalternative.aers.psu.edu).

5. Lanjabi-Sharahi M. (2009). Soil wetting pattern under subsurface drip irrigation with a geotextile envelope. M.Sc. thesis, Department of Water Engineering, Faculty of Agriculture, Shahrekord University, Shahrekord, Iran.

6. Mady, A. A., Mehanna, H. M., Okasha, E. M. (2012). Maximize water conservation of sandy soil by using Geotextile material under surface drip irrigation system. Journal of Applied Sciences Research, 8(8), 4014–4022.

7. Page, A. L., Miller, R. H., Keeny, D. R. (1982). Methods of Soil analysis, part ii chemical and microbiological properties. (2nd Ed), Amer. Soc. Agron. Monograph No. 9, Madison, Wisconsin. USA.

8. Reich D., R. Godin, Chávez, J. L., Broner, I. (2009). Subsurface drip irrigation. Colorado State University, U. S. Department of Agriculture, and Colorado counties cooperating.

9. Tabatabaei, S. H., Najafi, P. (2005). Effect of treated municipal wastewater injection with different irrigation methods in soil quality parameters, Proceedings of the 9th international conference on environmental sciences and technology, Rhodes Island, Greece.

10. Van Wambeke, A. (1992). Soils of the tropics – Properties and appraisal. McGraw Hill, Inc., New York.

CHAPTER 13

DESIGN OF A CORN PLANTER FOR SUBSURFACE MICRO IRRIGATION

M. M. MORAD, M. KH. AFIFI, H. M. HEKAL, and E. E. ABD-EL AATY

CONTENTS

Modified from Morad, M.M., M. Kh. Afifi, H. M. Hekal, and E. E. Abd-El-Aaty (2010). "Development of a planting machine used under sub surface irrigation system," *Misr Journal of Agricultural Engineering* 27(1), 2010. Modified by the authors. Originally published via open access.

In this chapter: 1 feddan = 0.42 hectares = 60×70 meter = 4200 m² = 1.038 acres = 24 kirat. A feddan (Arabic) is a unit of area. It is used in Egypt, Sudan, and Syria. The feddan is not an SI unit and in Classical Arabic, the word means 'a yoke of oxen': implying the area of ground that can be tilled in a certain time. In Egypt the feddan is the only nonmetric unit, which remained in use following the switch to the metric system. A feddan is divided into 24 Kirats (175 m²). In Syria, the feddan ranges from 2295 square meters (m²) to 3443 square meters (m²).

In this chapter: 1 L.E. = 0.14 US$. The Egyptian pound (Arabic: جنيه مصرى'*Genēh Maṣri;* E£ or ج.م; code: EGP) is the currency of Egypt. It is divided into 100 piastres, or ersh [قرش; *ʔerʃ; plural* شورق; German: *Groschen*], or 1,000 millimes [Arabic:ˈmælˈliːm; French: *Millime*]. The ISO 4217 code is EGP. Locally, the abbreviation LE or L.E., which stands for *livre égyptienne* [French for Egyptian pound] is frequently used. E£ and £E are rarely used. The name *Genēh/Geni* [geˈneː(h), ˈgeni] is derived from the Guinea coin, which had almost the same value of 100 piastres at the end of the nineteenth century.

13.1   INTRODUCTION

It is evident, that the increase of any crop production in both quantity and quality does not depend only on the improvement of soil and plant conditions, but also largely on using new irrigation systems as well as using improved methods and technology to fulfill the agricultural operations in the correct time to keep down the cost of production. The objectives of irrigation management should be shifted from obtaining maximum grain yield per cultivated area to maximum grain yield per unit of water. In other words, the best irrigation system is the system by which water use efficiency (WUE) is maximized. The matching of the water irrigation system to the agricultural machines is considered an important question that must be answered.

From this point of view surface drip irrigation system (SDIS), agricultural machines can be used after separating lateral parts and installing them after finishing the agricultural operations. On the contrary in subsurface drip irrigation system (SSDIS) despite of its advantages, it is difficult to use the agricultural machines because lateral lines in the irrigation system are installed underground. Therefore, care must be taken to design, construct, develop and operate special machines to be suitable for use under SSDIS, taking into consideration both factors related to irrigation and efficiencies of agricultural machines.

Bucks [4] stated that practical experience along with education and training had helped to improve the design and management of micro irrigation systems. Emitter clogging continued to be the nasty problem. However, the use of better filtration and chemical treatment systems had reduced or solved most of the problems. Crops grown under SSDIS must give equal or higher yield compared to those grown under SDIS. Abdel-Rahman [1] mentioned that the SSDIS is superior to the SDIS because of:

- System parts are protected from sunlight (ultra violet ray) that increases the parts life.
- The decrease of the moisture on surface layer resulting in reduction of attack of crops by insects and diseases.
- The SSDIS decreases the soil evaporation with increased water use efficiency along with fertilizing over subsurface trickle irrigation.

    - The water use efficiency increases in subsurface irrigation.
    - Using subsurface irrigation produces good moisture distribution in the soil profile.
    - Fertigation reduces labor cost and saves amount of fertilizers/ha.

Karayel and Ozmerzi [7] stated that variability in the seed spacing with a precision vacuum seeder was increased with increasing in forward speed. They revealed that forward speed of 1 m/s consistently produced a better seed pattern than 1.5 and 2 m/s for precision sowing of melon and cucumber seeds. Gomaa [5] noticed that forward speed of a planter had a significant effect on the field capacity and they observed greatest value of field capacity of about 3.68 fed/h. at 8.0 km/h forward speed for mechanical planter, while it was 3.52 fed/h at the same planting forward speed for pneumatic planter. Also, the results showed that the planting forward speed had an inverse effect on field efficiency. Gomaa also added that the pneumatic planter gave the highest values of germination ratio of 90.7% compared to 85.7% in case of mechanical planter both at forward speed of 3.16 km/h. Meanwhile the lowest values of total seed losses with mechanical planter were observed under the different levels of planting forward speeds.

Hanson and May [6] reported that subsurface drip irrigation in fine-textured salt-affected soils can increase yield and profit of tomatoes compared to sprinkler irrigation with acceptable levels of soluble solids (mainly due to the soil salinity at these locations). Drip irrigation also can control subsurface drainage to the shallow ground water.

The objectives of this research study were to:

1. Design and develop a small scale, self-propelled corn planter to be suitable in lands irrigated with subsurface drip irrigation.
2. Determine primary irrigation parameters to improve the uniformity of water distribution.
3. Optimize planter operating parameters to improve uniformity of seed distribution.
4. Evaluate the developed machine by comparing it to the manual planting, on the basis of cost economics.

13.2 MATERIALS AND METHODS

During the corn (cv. Triple Hybrid) growing season of 2008, Field experiments were carried out at Ras Sudr Research Station, South Sinai Governorate, Egypt. Based on the mechanical analysis, the soil was classified as sandy loam (Table 1). The experimental field was about 800 m², divided into two similar plots (each of 400 m², 25×16 m). The first plot was planted manually while the second plot was planted mechanically using the developed planter. In the two plots, a subsurface drip irrigation system was installed before planting.

TABLE 13.1 Some Soil Physical Properties at Experimental Site

Soil Depth (cm)	Particle size, %				Soil texture	Bulk Density g/cm³	FC %	PWP %	AWC %
	Coarse sand	Fine sand	Silt	Clay					
0–20	3.76	65.9	17.14	13.18	SL	1.45	35.12	22.33	12.79
20–40	2.17	68.1	14.63	15.07	SL	1.64	20.26	6.99	13.27

SL = Sandy loam; PWP = Permanent wilting percentage; FC = Field capacity,
AWC = FC – PWP

13.2.1 LIST OF EQUIPMENTS AND MATERIALS

- Tractor (New Holland) of 80 HP (58.8 kW).
- Rotary cultivator made in Turkey 14 HP (9.3KW).
- Components of **subsurface drip irrigation system** (Fig. 13.1): Mainline PE 75 mm diameter, submain line 50 mm, lateral line 16 mm with GR two emitters/m, lateral line 16 mm with GR 3 emitters/m, Stopcock 75 mm/3," Stopcock 50 mm/2," Tee 16 mm, Ends line 75 mm, Ends line 50 mm, and Ends line 16 mm.

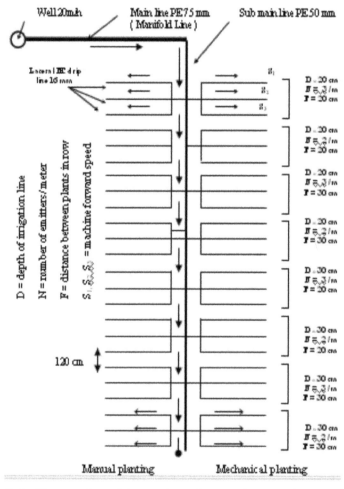

FIGURE 13.1 The experimental layout for evaluation of a corn planter and the drip irrigation system.

13.2.2 MEASUREMENTS

Following indicators were considered to evaluate the developed planter and the sub-surface drip irrigation system.

1. Moisture content: Measuring of moisture content in every treatment is considered very important to achieve uniformity of water distribution. It is calculated by the following formula:

$$MC\ (Wb) = 100 \times \{[Ww - Wd] / [Ww]\} \qquad (1)$$

where: MC = soil moisture content, %; Wb = Wet basis; Ww = wet soil mass, g; and Wd = dry soil mass, g.

2. Water distribution uniformity (CU%) is calculated as follows:

$$CU\ \% = 1 - \frac{\sum (X - \overline{X})}{\overline{X}n} \qquad (2)$$

where: CU % = water distribution uniformity; × = Moisture content values in roots revealing area of 20 cm; $\overline{X}$ = average of moisture content values; and n = number of soil samples.

3. Emergence ratio: The emergence period was estimated by noticing daily the number of seedlings, which were emerged. When the seedling emergence was grown, the emergence ratio (Er) was estimated by the following equation:

$$= \sqrt{\frac{\sum (x - \mu)^2}{N}} \qquad (3)$$

where: Er = Emergence ratio; NP = Average number of seedlings per unit area; and NS = Average number of delivered seeds in the same area.

4. Plant characteristics: Following characteristics were investigated during flowering and harvesting period:

    a. Average number of plants in furrow.
    b. Average plant height in cm (measured from soil surface to the top of main stem).
    c. Average plant diameter.
    d. Average mass of seeds per plant (calculated as an average of ten plants in grams for each treatment).

5. Crop yield: The corn yield was determined for manual and mechanical planting methods. A number of samples along the row was taken from different locations in each treatment at random, and then were weighted and integrated to determine the average yield of corn per feddan.

6. Longitudinal and transverse dispersion: The dispersion of plants about the center of row is determined according to the following formula [8].

$$= \sqrt{\frac{\sum (x - \mu)^2}{N}} \tag{4}$$

where:$\sum (x - \mu)^2$ = The sum of squares of variance of seed scattering; and N = The number of hills.

7. Experimental conditions were: In both manual and mechanical planting, the distance between the two rows was 60 cm while the planting depth was about 5 cm. The subsurface drip irrigation system was studied under the following variables:

- Two depths of drip irrigation lines (20 and 30 cm).
- Two number of emitters per meter (2/m and 3/m).
- The developed planting machine was studied under the following variables:
  - Three different forward speeds (2.5, 3.2 and 3.88 km/h).
  - Two distances between plants in the same row (20 and 30 cm).

13.3   DEVELOPMENT OF A CORN PLANTER

A rotary cultivator was modified into a self-propelled planter, suitable for farms with subsurface drip irrigation system (SSDIS). The developed machine is shown in Fig. 13.2. The developed machine consisted of the following parts:

1. A rotary cultivator: Self-propelled with an engine of 14 HP (9 KW), gear box with three forward speeds and one reverse speed, and two wheels of 14×75 size, weight 380 kg.
2. Hitching point: It is fabricated to join planting unit with the machine frame by small buckles from iron shaft with 6 mm of thickness and 120 cm of length.
3. The two planting units: Two planting units (John Deere) with length 100 cm and mass of 120 kg were installed on the rotary cultivator. The components of the planting units were as following:

- Grain hopper: A small and simple grain hopper was built from light iron with a capacity of 10 kg. Grain hopper was provided with an agitator to prevent seeds from shaping a hill above feeding holes. Grain hopper was of a cylinder shape.
- Feeding cells: Two feeding cells were fabricated, so as to give two distances between plants in the same row during the planting operation.
- Disk opener: Two disk openers for the two planting units were fabricated from strong iron steel to penetrate soil and reform seeds flutes.
- Fluted wheel: One wheel in every planting unit was used as covering wheel to cover the seeds and other for transmitting motion to feeding cells through sprockets.

• Transmission system: Motion was transmitted from a fluted wheel to the planting units by means of sprockets and chains.

13.3.1 PARAMETERS TO EVALUATE THE CORN PLANTER

1. Ground wheel slip: Slippage percentage was calculated by using the following equation [2]:

$$\text{Slippage, \%} = 100 \times \{[D_{noload} - D_{load}]/[D_{load}]\} \qquad (5)$$

where: D_{noload} = distance traveled by the machine with no load; and D_{load} = distance traveled by the machine with load.

2. Field capacity and field efficiency:

$$\text{FE, \%} = 100 \times [(EFC)/(TFC)] \qquad (6)$$

where: FE = field efficiency of the machine, %. Effective field capacity (EFC, fed./h) is the actual average working rate of area and the theoretical field capacity (TFC, fed./h) is calculated by multiplying machine forward speed by the effective working width of the machine.

3. Fuel consumption was recorded by accurately measuring the decrease in fuel level in tank immediately after executing each operation.

4. The required power (kW) for planter was calculated according to Barger et al. [3]:

$$\text{Power, kW} = F \times \rho_f \times C_v \times \tau_{th} \times \tau_{m} \times (427/75) \times (1/1.36) \qquad (7)$$

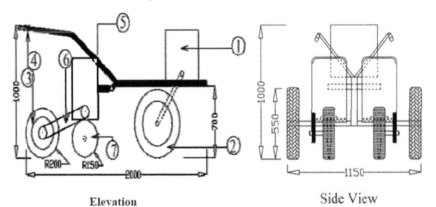

Elevation   Side View

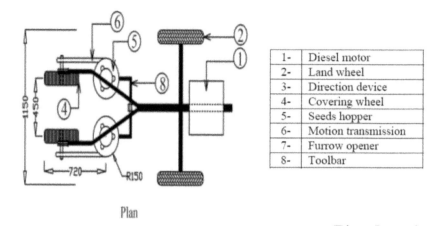

1-	Diesel motor
2-	Land wheel
3-	Direction device
4-	Covering wheel
5-	Seeds hopper
6-	Motion transmission
7-	Furrow opener
8-	Toolbar

Plan

(Dimn. In, mm)

FIGURE 13.2   Schematic diagram of the corn planter, developed for use in SSDIS.

where: F = fuel consumption, L/h; ρ_f = fuel density (0.85 kg/L for diesel fuel); C_v = lower calorific value of fuel = 104 k Cal/kg; τ_{th} = thermal efficiency of engine = 40%; τ_m = mechanical efficiency of engine = 80%; 427 = thermo-mechanical equivalent, kg. m/k Cal; and 75 = conversion factor = (HP)/(Kg.m/sec).

5.   Energy requirements (kW-h/fed.):

$$Energy = \frac{Power, kW}{Effective\ field\ capacity\ fed\ /\ h} \qquad (8)$$

6.   Machinery cost analysis: The machine cost is determined using the conventional method of estimating both fixed and variable costs.

7.   The planting cost (LE/fed.):

Planting cost = [(Machine cost, LE/h)/(Effective field capacity, fed/h)]   (9)

13.4   RESULTS AND DISCUSSION

13.4.1   PERFORMANCE OF SUBSURFACE DRIP IRRIGATION

Figure 13.3 indicates the effects of depth of drip line and number of emitters on soil moisture content. Soil moisture content was increased from 10.6% to 13.4% by increasing installation depth of drip line from 20 to 30 cm with 3 emitters per meter length of drip line. And increasing the drip line depth with two emitters/m caused increase in moisture content from 8.26% to 9.7%. This increase in soil moisture content is attributed to the decrease of water losses by evaporation and the effect of sun heat on soil surface. While the increase in soil moisture content by increasing number of emitters along the plant row is due to the increase in irrigation depth, discharge/m, intensification of moisture content in the soil, and reduction in leaching rate.

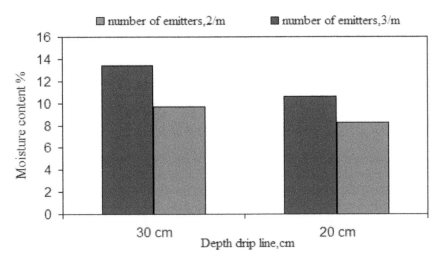

FIGURE 13.3 Effects of installation depth of drip line and number of emitters on soil moisture content.

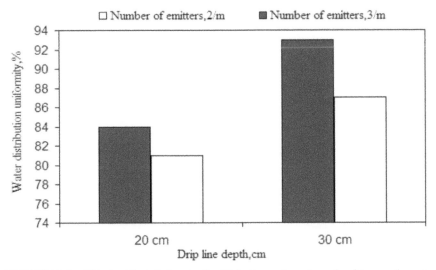

FIGURE 13.4 Effects of installation depth of drip line and number of emitters on the water distribution uniformity (CU, %).

Figure 13.4 shows the influence of depth of drip line and number of emitters/m on water distribution uniformity (CU%) in the field. Increasing installation depth of drip line from 20 cm to 30 cm with three emitters per meter of drip line increased water distribution uniformity from 84 to 93%, while the same increase in drip line depth with 2 emitters/m increased water distribution uniformity from 81% to 87%. The increase of both drip line depth and number of emitters increased water distribution uniformity due to the increase in retention time of water in root zone and decreasing water losses by evaporation.

13.4.2 PERFORMANCE OF THE DEVELOPED PLANTER

13.4.2.1 EFFECTS OF FORWARD SPEED ON FIELD CAPACITY, FIELD EFFICIENCY AND GROUND WHEEL SLIP

Representative values of field capacity, field efficiency and ground wheel slip versus machine forward speed during planting of corn are given in Fig. 13.5. Results show that there was an increase of field capacity from 0.58 to 0.91 fed./h and ground wheel slip from 4.7 to 9.33%; and decrease in field efficiency from 89.2 to 81%, when the machine forward speed was increased from 2.5 to 3.88 km/h. The reduction in field efficiency by increasing forward speed was due to the less theoretical time consumed in comparison with the other items. It is obvious that lower forward speed of 2.5 km/h tends to increase field capacity, but at the same time significantly decrease was observed in field capacity and the vice versa was noticed with the forward speed of 3.88 km/h. Therefore, forward speed 3.2 km/h is considered an optimum value for the planting operation.

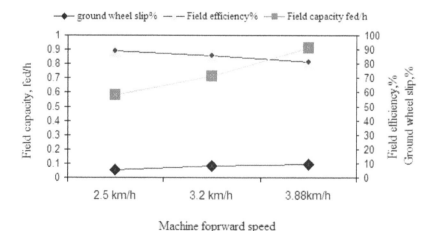

FIGURE 13.5   Effects of forward speed on field capacity, field efficiency and ground wheel slip.

13.4.2.2   EFFECTS OF FORWARD SPEED ON SEED DISPERSION

The effects of forward speed on longitudinal and transverse dispersion of seed for two plant-to-plant spacing within the row are presented in Fig. 13.6. When the forward speed was increased from 2.5 to 3.88 km/h, there was an increase in the longitudinal dispersion of seed from 2.47 to 2.88 cm and from 2.66 to 3 cm for 20 and 30 cm plant-to-plant spacing's, respectively. Also, transverse dispersion was increased from 8 to 12.12 cm and from 9.46 to 12.37 cm in these spacing's. Generally, high forward speeds affect negatively on uniformity of seed dispersion, because increasing forward speed led to accelerate the machine and vibrate it consequently, and irregular hills and irregular distances are occurred.

13.4.2.3   EFFECTS OF FORWARD SPEED ON EMERGENCE RATIO

The Fig. 13.7 shows that machine forward speed and emergence ratio are inversely related. Increasing forward speed from 2.5 to 3.88 km/h decreased emergence ratio from 92 to 88.2% and from 89 to 83.7%, for 20 and 30 cm plant to plant spacing down the row, respectively. The decrease in emergence ratio by increasing forward speed is attributed to high increase in seed dispersion under high machine speeds resulting in low emergence ratio.

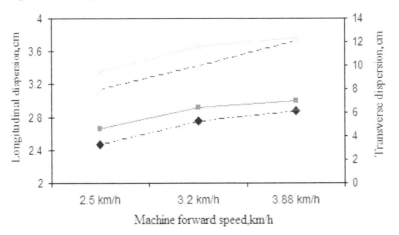

- - - - transverse dispersion with 20cm distance between plants
———— transverse dispersion with 30cm distance between plants
- ·◆-· longitudinal dispersion with 20cm distance between plants
——■—— longitudinal dispersion with 30cm distance between plants

FIGURE 13.6   Effects of forward speed on longitudinal and transverse seed dispersion for plant to plant spacing's.

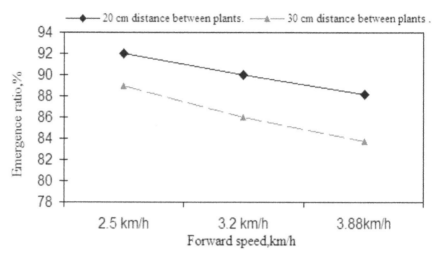

FIGURE 13.7   Effects of forward speed on corn emergence ratio.

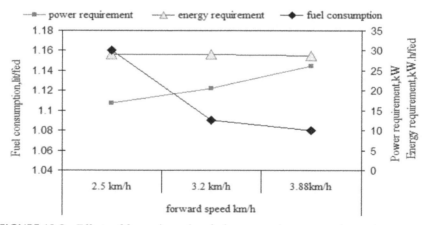

FIGURE 13.8   Effects of forward speed on fuel, power and energy requirements.

TABLE 13.2   Effects of Planting Methods on Plant Characteristics and Corn Yield

Factors	Mechanical planting	Manual planting
Plant height, cm	135	118
Plant diameter, cm	2.5	1.8
Number of seeds per plant, seeds	451	374
Crop yield, Mg/feddans	3.31	2.94

13.4.2.4   EFFECTS OF FORWARD SPEED ON FUEL, POWER AND ENERGY REQUIREMENTS

The Fig. 13.8 show that both fuel and energy requirements were decreased, when the forward speed was increased. The vice versa was observed with the decreased fuel consumption from 1.16 to 1.08 L/fed and also decreased energy requirements from 29.1 to 28.79 kW-h/fed. The required power increased from 16.88 to 26.2 kW. The decrease of fuel and energy by increasing forward speed is attributed to the increase in field capacity, resulting in low values of fuel and energy per feddan.

13.4.3   COMPARISON BETWEEN MANUAL AND MECHANICAL PLANTING OF CORN

13.4.3.1   EFFECTS OF PLANTING METHODS ON PLANT CHARACTERISTICS AND CORN YIELD

Table 2 indicates effects of two planting methods on plant characteristics and corn yield: plant height, plant diameter, number of seeds per plant, and crop yield. Observed data shows that the mechanical planting method gave best values of all these parameters comparing to manual planting. Mechanical planting gave good emergence ratio, good seed distribution, good seed depth and healthy corn seedlings. This attributed to higher yields with the use of mechanical planting.

13.4.3.2   EFFECTS OF PLANTING METHOD ON PLANTING COSTS

The cost of seed planting with a corn planter in this chapter was 20.2, 26 and 30.2 LE/fed for forward speed of 2.5, 3.2 and 3.88 km/h, respectively, compared to 72 LE/fed with a manual planting by using one labor per feddan. The decrease in planting cost by increasing forward speed is attributed to the increase in machine field capacity. While the increase in planting cost with the use of manual planting is due to the very low working output of labor.

13.5   CONCLUSIONS

- Subsurface drip irrigation system achieved the best results with an installation depth of 30 cm and 3 emitters per meter length of drip line.
- The developed corn planter gave high field efficiency at a forward speed of 3.2 km/h and seed to seed spacing of 30 cm.

13.6 SUMMARY

A Self-propelled corn planter was designed and developed to be suitable for farms under subsurface drip irrigation system. The developed machine was evaluated in the corn field to study effects of two installation depths of drip lines, two emitter spacing's, three forward speeds, and two plant to plant spacing's down the row. The subsurface drip irrigation system was evaluated for variation in soil moisture content and water distribution uniformity. The developed corn planter was evaluated by comparing its performance to the manual method in terms of: Emergence ratio, longitudinal dispersion, power and energy requirements, crop yield and seed planting cost. The experimental results for subsurface drip irrigation concluded that:

- Lateral line depth of 30 cm with three emitters per meter was considered optimum to achieve high uniformity of water distribution and high water use efficiency.
- Machine forward speed of 3.2 km/h with 30 cm plant-to-plant spacing gave high crop yield and low planting cost.

KEYWORDS

- corn
- corn planter
- crop yield
- Desert Research Center
- drip irrigation
- drip line
- emergence ratio
- emitter
- energy requirements
- field capacity
- field efficiency
- forward speed
- lateral line
- longitudinal dispersion
- micro irrigation
- North Sinai
- plant to plant spacing

- planter
- planting cost
- seed flute
- soil moisture
- subsurface drip irrigation
- transverse dispersion
- water distribution uniformity, CU
- water use efficiency, WUE

REFERENCES

1. Abdel-Rahman, G. A. (1996). The effect of applying drip irrigation system under certain environmental resources on soil productivity deterioration at North Sinai. *Miser J. Agric. Eng.*, 13(2), 612–624.
2. Awady, M. N. (1992). *Farm machines*. College of Agric., Ain Shams Univ., 120 pages.
3. Barger, E. L., Bainer, R. (1963). *Tractors and their power units*. John Wiley & Sons Inc., New York.
4. Bucks, D. A. (1995). Historical developments in micro irrigation. Proceeding of the 5th International Micro Irrigation Congress. April 2–6, Orlando, Florida. ASAE, 2950 Niles Rd., St. Joseph, MI 49085–9659. Pages 1–5.
5. Gomaa, S. M. (2003). Performance evaluation of pneumatic and mechanical planters for cowpea planting. *Misr J. Agric. Eng.*, 20(4), 965–979.
6. Hanson, B., May, D. (2004). Effect of subsurface drip irrigation on processing tomato yield, water table depth, soil salinity, and profitability. *Agricultural Water Management*, 68(1), 1–17.
7. Karayel, D., Ozmerzi, A. (2004). Effect of forward speed and seed spacing uniformity on a precision vacuum metering unit for melon and cucumber seeds. *HortTechnology*, 14(3), 364–367.
8. Stell, R. G., Torrie, S. H. (1980). *Principles and procedures of statistics*. McGraw Hill Company, NY, USA.

APPENDICES

(Modified and reprinted with permission from: Goyal, Megh R., 2012. Appendices. Pages 317–332. In: *Management of Drip/Trickle or Micro Irrigation* edited by Megh R. Goyal. New Jersey, USA: Apple Academic Press Inc.)

APPENDIX A

CONVERSION SI AND NON-SI UNITS

To convert the Column 1 in the Column 2, Multiply by	Column 1 Unit SI	Column 2 Unit Non-SI	To convert the Column 2 in the Column 1 Multiply by

LINEAR

0.621 ——— kilometer, km (10^3 m)		miles, mi ———————	1.609
1.094 ——— meter, m		yard, yd ———————	0.914
3.28 ——— meter, m		feet, ft ———————	0.304
3.94×10^{-2}—— millimeter, mm (10^{-3})		inch, in ———————	25.4

SQUARES

2.47—— hectare, he		acre ———————	0.405
2.47—— square kilometer, km^2		acre ———————	4.05×10^{-3}
0.386—— square kilometer, km^2		square mile, mi^2———	2.590
2.47×10^{-4} —— square meter, m^2		acre ———————	4.05×10^{-3}
10.76—— square meter, m^2		square feet, ft^2———	9.29×10^{-2}
1.55×10^{-3}—— mm^2		square inch, in^2———	645

CUBICS

9.73×10^{-3}—— cubic meter, m^3		inch-acre ———————	102.8
35.3—— cubic meter, m^3		cubic-feet, ft^3———	2.83×10^{-2}
6.10×10^4—— cubic meter, m^3		cubic inch, in^3———	1.64×10^{-5}
2.84×10^{-2}—— liter, L (10^{-3} m^3)		bushel, bu ———————	35.24

1.057 —— liter, L	liquid quarts, qt ————— 0.946
3.53×10^{-2}—— liter, L	cubic feet, ft^3 ————— 28.3
0.265 —— liter, L	gallon ————— 3.78
33.78 —— liter, L	fluid ounce, oz ————— 2.96×10^{-2}
2.11 —— liter, L	fluid dot, dt ————— 0.473

WEIGHT

2.20×10^{-3} —— gram, g (10^{-3} kg)	pound, ————— 454
3.52×10^{-2} —— gram, g (10^{-3} kg)	ounce, oz ————— 28.4
2.205 —— kilogram, kg	pound, lb ————— 0.454
10^{-2} —— kilogram, kg	quintal (metric), q ——— 100
1.10×10^{-3} —— kilogram, kg	ton (2000 lbs), ton ——— 907
1.102 —— mega gram, mg	ton (US), ton ————— 0.907
1.102 —— metric ton, t	ton (US), ton ————— 0.907

YIELD AND RATE

0.893 —— kilogram per hectare	pound per acre ————— 1.12
7.77×10^{-2}— kilogram per cubic meter	pound per fanega ——— 12.87
1.49×10^{-2}— kilogram per hectare	pound per acre, 60 lb —— 67.19
1.59×10^{-2}— kilogram per hectare	pound per acre, 56 lb —— 62.71
1.86×10^{-2}— kilogram per hectare	pound per acre, 48 lb —— 53.75
0.107 —— liter per hectare	galloon per acre ——— 9.35
893 —— ton per hectare	pound per acre ————— 1.12×10^{-3}
893 —— mega gram per hectare	pound per acre ————— 1.12×10^{-3}
0.446 —— ton per hectare	ton (2000 lb) per acre —— 2.24
2.24 —— meter per second	mile per hour ————— 0.447

SPECIFIC SURFACE

10 —— square meter per kilogram	square centimeter per gram ————— 0.1
10^3 —— square meter per kilogram	square millimeter per gram ————— 10^{-3}

PRESSURE

9.90 ——— megapascal, MPa	atmosphere ——————— 0.101
10 ——— megapascal	bar ——————————— 0.1
1.0 ——— megagram per cubic meter	gram per cubic centimeter ——————— 1.00
2.09×10^{-2} — pascal, Pa	pound per square feet ——— 47.9
1.45×10^{-4} — pascal, Pa	pound per square inch ——— 6.90×10^3

To convert the Column 1 in the Column 2,	Column 1	Column 2	To convert the Column 2 in the Column 1
	Unit	Unit	
Multiply by	SI	Non-SI	Multiply by

TEMPERATURE

1.00 (K-273) — Kelvin, K	centigrade, °C ——— 1.00 (C+273)
(1.8 C + 32) — centigrade, °C	Fahrenheit,°F ——— (F--32)/1.8

ENERGY

9.52×10^{-4} —Joule J	BTU —————— 1.05×10^3
0.239 ———	Joule, J	calories, cal ——— 4.19
0.735 ———	Joule, J	feet-pound ——— 1.36
2.387×10^5 —	Joule per square meter	calories per square centimeter ——— 4.19×10^4
10^5———	Newton, N	dynes ——— 10^{-5}

WATER REQUIREMENTS

9.73×10^{-3} ——— cubic meter	inch acre ——— 102.8
9.81×10^{-3} ——— cubic meter per hour	cubic feet per second ——— 101.9
4.40 ————— cubic meter per hour	galloon (US) per minute ——— 0.227
8.11 ————— hectare-meter	acre-feet ——— 0.123
97.28 ————— hectare-meter	acre-inch ——— 1.03×10^{-2}
8.1×10^{-2} ——— hectare centimeter	acre-feet ——— 12.33

CONCENTRATION

1 ———————	centimol per kilogram	milliequivalents per 100 grams ——————— 1
0.1 —————	gram per kilogram	percents ———————————— 10
1 —————	milligram per kilogram	parts per million ————— 1

NUTRIENTS FOR PLANTS

2.29 ——— P	P_2O_5 ——————————— 0.437
1.20 ——— K	K_2O ——————————— 0.830
1.39 ——— Ca	CaO ——————————— 0.715
1.66 ——— Mg	MgO ——————————— 0.602

NUTRIENT EQUIVALENTS

Column A	Column B	Conversion A to B	Equivalent B to A
N	NH_3	1.216	0.822
	NO_3	4.429	0.226
	KNO_3	7.221	0.1385
	$Ca(NO_3)_2$	5.861	0.171
	$(NH_4)_2SO_4$	4.721	0.212
	NH_4NO_3	5.718	0.175
	$(NH_4)_2HPO_4$	4.718	0.212
P	P_2O_5	2.292	0.436
	PO_4	3.066	0.326
	KH_2PO_4	4.394	0.228
	$(NH_4)_2HPO_4$	4.255	0.235
	H_3PO_4	3.164	0.316
K	K_2O	1.205	0.83
	KNO_3	2.586	0.387
	KH_2PO_4	3.481	0.287
	Kcl	1.907	0.524
	K_2SO_4	2.229	0.449

Ca	CaO	1.399	0.715
	$Ca(NO_3)_2$	4.094	0.244
	$CaCl_2 \times 6H_2O$	5.467	0.183
	$CaSO_4 \times 2H_2O$	4.296	0.233
Mg	MgO	1.658	0.603
	$MgSO_4 \times 7H_2O$	1.014	0.0986
S	H_2SO_4	3.059	0.327
	$(NH_4)_2SO_4$	4.124	0.2425
	K_2SO_4	5.437	0.184
	$MgSO_4 \times 7H_2O$	7.689	0.13
	$CaSO_4 \times 2H_2O$	5.371	0.186

APPENDIX B

PIPE AND CONDUIT FLOW

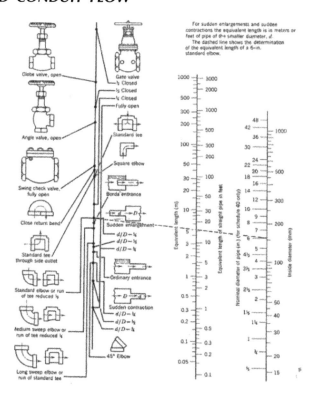

APPENDIX C

PERCENTAGE OF DAILY SUNSHINE HOURS: FOR NORTH AND SOUTH HEMISPHERES

Latitude	Jan	Feb	Mar	Apr	May	Jun	Jul	Aug	Sep	Oct	Nov	Dec
NORTH												
0	8.50	7.66	8.49	8.21	8.50	8.22	8.50	8.49	8.21	8.50	8.22	8.50
5	8.32	7.57	8.47	3.29	8.65	8.41	8.67	8.60	8.23	8.42	8.07	8.30
10	8.13	7.47	8.45	8.37	8.81	8.60	8.86	8.71	8.25	8.34	7.91	8.10
15	7.94	7.36	8.43	8.44	8.98	8.80	9.05	8.83	8.28	8.20	7.75	7.88
20	7.74	7.25	8.41	8.52	9.15	9.00	9.25	8.96	8.30	8.18	7.58	7.66
25	7.53	7.14	8.39	8.61	9.33	9.23	9.45	9.09	8.32	8.09	7.40	7.52
30	7.30	7.03	8.38	8.71	9.53	9.49	9.67	9.22	8.33	7.99	7.19	7.15
32	7.20	6.97	8.37	8.76	9.62	9.59	9.77	9.27	8.34	7.95	7.11	7.05
34	7.10	6.91	8.36	8.80	9.72	9.70	9.88	9.33	8.36	7.90	7.02	6.92
36	6.99	6.85	8.35	8.85	9.82	9.82	9.99	9.40	8.37	7.85	6.92	6.79
38	6.87	6.79	8.34	8.90	9.92	9.95	10.1	9.47	3.38	7.80	6.82	6.66
40	6.76	6.72	8.33	8.95	10.0	10.1	10.2	9.54	8.39	7.75	6.72	7.52
42	6.63	6.65	8.31	9.00	10.1	10.2	10.4	9.62	8.40	7.69	6.62	6.37
44	6.49	6.58	8.30	9.06	10.3	10.4	10.5	9.70	8.41	7.63	6.49	6.21
46	6.34	6.50	8.29	9.12	10.4	10.5	10.6	9.79	8.42	7.57	6.36	6.04
48	6.17	6.41	8.27	9.18	10.5	10.7	10.8	9.89	8.44	7.51	6.23	5.86
50	5.98	6.30	8.24	9.24	10.7	10.9	11.0	10.0	8.35	7.45	6.10	5.64
52	5.77	6.19	8.21	9.29	10.9	11.1	11.2	10.1	8.49	7.39	5.93	5.43
54	5.55	6.08	8.18	9.36	11.0	11.4	11.4	10.3	8.51	7.20	5.74	5.18
56	5.30	5.95	8.15	9.45	11.2	11.7	11.6	10.4	8.53	7.21	5.54	4.89
58	5.01	5.81	8.12	9.55	11.5	12.0	12.0	10.6	8.55	7.10	4.31	4.56
60	4.67	5.65	8.08	9.65	11.7	12.4	12.3	10.7	8.57	6.98	5.04	4.22
SOUTH												
0	8.50	7.66	8.49	8.21	8.50	8.22	8.50	8.49	8.21	8.50	8.22	8.50
5	8.68	7.76	8.51	8.15	8.34	8.05	8.33	8.38	8.19	8.56	8.37	8.68
10	8.86	7.87	8.53	8.09	8.18	7.86	8.14	8.27	8.17	8.62	8.53	8.88
15	9.05	7.98	8.55	8.02	8.02	7.65	7.95	8.15	8.15	8.68	8.70	9.10
20	9.24	8.09	8.57	7.94	7.85	7.43	7.76	8.03	8.13	8.76	8.87	9.33

25	9.46	8.21	8.60	7.74	7.66	7.20	7.54	7.90	8.11	8.86	9.04	9.58
30	9.70	8.33	8.62	7.73	7.45	6.96	7.31	7.76	8.07	8.97	9.24	9.85
32	9.81	8.39	8.63	7.69	7.36	6.85	7.21	7.70	8.06	9.01	9.33	9.96
34	9.92	8.45	8.64	7.64	7.27	6.74	7.10	7.63	8.05	9.06	9.42	10.1
36	10.0	8.51	8.65	7.59	7.18	6.62	6.99	7.56	8.04	9.11	9.35	10.2
38	10.2	8.57	8.66	7.54	7.08	6.50	6.87	7.49	8.03	9.16	9.61	10.3
40	10.3	8.63	8.67	7.49	6.97	6.37	6.76	7.41	8.02	9.21	9.71	10.5
42	10.4	8.70	8.68	7.44	6.85	6.23	6.64	7.33	8.01	9.26	9.8	10.6
44	10.5	8.78	8.69	7.38	6.73	6.08	6.51	7.25	7.99	9.31	9.94	10.8
46	10.7	8.86	8.90	7.32	6.61	5.92	6.37	7.16	7.96	9.37	10.1	11.0

APPENDIX D

PSYCHOMETRIC CONSTANT (Γ) FOR DIFFERENTAL TITUDES (Z)

$$\gamma = 10^{-3} \, [(C_p.P) \div (\varepsilon.\lambda)] = (0.00163) \times [P \div \lambda]$$

γ, psychrometric constant [kPa C^{-1}]
c_p, specific heat of moist air = 1.013
[kJ kg^{-1}^{0}C^{-1}]
P, atmospheric pressure [kPa].

ε, ratio molecular weight of water
vapor/dry air = 0.622
λ, latent heat of vaporization [MJ kg^{-1}]
= 2.45 MJ kg^{-1} at 20°C.

Z (m)	γ kPa/°C	z (m)	γ kPa/°C	z (m)	γ kPa/°C	z (m)	γ kPa/°C
0	0.067	1000	0.060	2000	0.053	3000	0.047
100	0.067	1100	0.059	2100	0.052	3100	0.046
200	0.066	1200	0.058	2200	0.052	3200	0.046
300	0.065	1300	0.058	2300	0.051	3300	0.045
400	0.064	1400	0.057	2400	0.051	3400	0.045
500	0.064	1500	0.056	2500	0.050	3500	0.044
600	0.063	1600	0.056	2600	0.049	3600	0.043
700	0.062	1700	0.055	2700	0.049	3700	0.043
800	0.061	1800	0.054	2800	0.048	3800	0.042
900	0.061	1900	0.054	2900	0.047	3900	0.042
1000	0.060	2000	0.053	3000	0.047	4000	0.041

APPENDIX E

SATURATION VAPOR PRESSURE [E_s] FOR DIFFERENTTEMPERATURES (T)

| \multicolumn{8}{c}{Vapor pressure function = e_s = [0.6108]*exp\{[17.27*T]/[T + 237.3]\}} |

T °C	e_s kPa	T °C	e_s kPa	T °C	e_s kPa	T °C	e_s kPa
1.0	0.657	13.0	1.498	25.0	3.168	37.0	6.275
1.5	0.681	13.5	1.547	25.5	3.263	37.5	6.448
2.0	0.706	14.0	1.599	26.0	3.361	38.0	6.625
2.5	0.731	14.5	1.651	26.5	3.462	38.5	6.806
3.0	0.758	15.0	1.705	27.0	3.565	39.0	6.991
3.5	0.785	15.5	1.761	27.5	3.671	39.5	7.181
4.0	0.813	16.0	1.818	28.0	3.780	40.0	7.376
4.5	0.842	16.5	1.877	28.5	3.891	40.5	7.574
5.0	0.872	17.0	1.938	29.0	4.006	41.0	7.778
5.5	0.903	17.5	2.000	29.5	4.123	41.5	7.986
6.0	0.935	18.0	2.064	30.0	4.243	42.0	8.199
6.5	0.968	18.5	2.130	30.5	4.366	42.5	8.417
7.0	1.002	19.0	2.197	31.0	4.493	43.0	8.640
7.5	1.037	19.5	2.267	31.5	4.622	43.5	8.867
8.0	1.073	20.0	2.338	32.0	4.755	44.0	9.101
8.5	1.110	20.5	2.412	32.5	4.891	44.5	9.339
9.0	1.148	21.0	2.487	33.0	5.030	45.0	9.582
9.5	1.187	21.5	2.564	33.5	5.173	45.5	9.832
10.0	1.228	22.0	2.644	34.0	5.319	46.0	10.086
10.5	1.270	22.5	2.726	34.5	5.469	46.5	10.347
11.0	1.313	23.0	2.809	35.0	5.623	47.0	10.613
11.5	1.357	23.5	2.896	35.5	5.780	47.5	10.885
12.0	1.403	24.0	2.984	36.0	5.941	48.0	11.163
12.5	1.449	24.5	3.075	36.5	6.106	48.5	11.447

APPENDIX F

SLOPE OF VAPOR PRESSURE CURVE (Δ) FOR DIFFERENTTEMPERATURES (T)

$$\Delta = [4098. \, e^0(T)] \div [T + 237.3]^2$$
$$= 2504\{\exp[(17.27T) \div (T + 237.2)]\} \div [T + 237.3]^2$$

T °C	Δ kPa/°C	T °C	Δ kPa/°C	T °C	Δ kPa/°C	T °C	Δ kPa/°C
1.0	0.047	13.0	0.098	25.0	0.189	37.0	0.342
1.5	0.049	13.5	0.101	25.5	0.194	37.5	0.350
2.0	0.050	14.0	0.104	26.0	0.199	38.0	0.358
2.5	0.052	14.5	0.107	26.5	0.204	38.5	0.367
3.0	0.054	15.0	0.110	27.0	0.209	39.0	0.375
3.5	0.055	15.5	0.113	27.5	0.215	39.5	0.384
4.0	0.057	16.0	0.116	28.0	0.220	40.0	0.393
4.5	0.059	16.5	0.119	28.5	0.226	40.5	0.402
5.0	0.061	17.0	0.123	29.0	0.231	41.0	0.412
5.5	0.063	17.5	0.126	29.5	0.237	41.5	0.421
6.0	0.065	18.0	0.130	30.0	0.243	42.0	0.431
6.5	0.067	18.5	0.133	30.5	0.249	42.5	0.441
7.0	0.069	19.0	0.137	31.0	0.256	43.0	0.451
7.5	0.071	19.5	0.141	31.5	0.262	43.5	0.461
8.0	0.073	20.0	0.145	32.0	0.269	44.0	0.471
8.5	0.075	20.5	0.149	32.5	0.275	44.5	0.482
9.0	0.078	21.0	0.153	33.0	0.282	45.0	0.493
9.5	0.080	21.5	0.157	33.5	0.289	45.5	0.504
10.0	0.082	22.0	0.161	34.0	0.296	46.0	0.515
10.5	0.085	22.5	0.165	34.5	0.303	46.5	0.526
11.0	0.087	23.0	0.170	35.0	0.311	47.0	0.538
11.5	0.090	23.5	0.174	35.5	0.318	47.5	0.550
12.0	0.092	24.0	0.179	36.0	0.326	48.0	0.562
12.5	0.095	24.5	0.184	36.5	0.334	48.5	0.574

APPENDIX G

NUMBER OF THE DAY IN THE YEAR (JULIAN DAY)

Day	Jan	Feb	Mar	Apr	May	Jun	Jul	Aug	Sep	Oct	Nov	Dec
1	1	32	60	91	121	152	182	213	244	274	305	335
2	2	33	61	92	122	153	183	214	245	275	306	336
3	3	34	62	93	123	154	184	215	246	276	307	337
4	4	35	63	94	124	155	185	216	247	277	308	338
5	5	36	64	95	125	156	186	217	248	278	309	339
6	6	37	65	96	126	157	187	218	249	279	310	340
7	7	38	66	97	127	158	188	219	250	280	311	341
8	8	39	67	98	128	159	189	220	251	281	312	342
9	9	40	68	99	129	160	190	221	252	282	313	343
10	10	41	69	100	130	161	191	222	253	283	314	344
11	11	42	70	101	131	162	192	223	254	284	315	345
12	12	43	71	102	132	163	193	224	255	285	316	346
13	13	44	72	103	133	164	194	225	256	286	317	347
14	14	45	73	104	134	165	195	226	257	287	318	348
15	15	46	74	105	135	166	196	227	258	288	319	349
16	16	47	75	106	136	167	197	228	259	289	320	350
17	17	48	76	107	137	168	198	229	260	290	321	351
18	18	49	77	108	138	169	199	230	261	291	322	352
19	19	50	78	109	139	170	200	231	262	292	323	353
20	20	51	79	110	140	171	201	232	263	293	324	354
21	21	52	80	111	141	172	202	233	264	294	325	355
22	22	53	81	112	142	173	203	234	265	295	326	356
23	23	54	82	113	143	174	204	235	266	296	327	357
24	24	55	83	114	144	175	205	236	267	297	328	358
25	25	56	84	115	145	176	206	237	268	298	329	359
26	26	57	85	116	146	177	207	238	269	299	330	360
27	27	58	86	117	147	178	208	239	270	300	331	361
28	28	59	87	118	148	179	209	240	271	301	332	362
29	29	(60)	88	119	149	180	210	241	272	302	333	363
30	30	—	89	120	150	181	211	242	273	303	334	364
31	31	—	90	—	151	—	212	243	—	304	—	365

APPENDIX H

STEFAN-BOLTZMANN LAW AT DIFFERENT TEMPERATURES (T):

$[\sigma*(T_K)^4] = [4.903 \times 10^{-9}]$, MJ K^{-4} m^{-2} day^{-1}
Where: $T_K = \{T[°C] + 273.16\}$

T	$\sigma*(T_K)^4$	T	$\sigma*(T_K)^4$	T	$\sigma*(T_K)^4$
Units					
°C	MJ m^{-2} d^{-1}	°C	MJ m^{-2} d^{-1}	°C	MJ m^{-2} d^{-1}
1.0	27.70	17.0	34.75	33.0	43.08
1.5	27.90	17.5	34.99	33.5	43.36
2.0	28.11	18.0	35.24	34.0	43.64
2.5	28.31	18.5	35.48	34.5	43.93
3.0	28.52	19.0	35.72	35.0	44.21
3.5	28.72	19.5	35.97	35.5	44.50
4.0	28.93	20.0	36.21	36.0	44.79
4.5	29.14	20.5	36.46	36.5	45.08
5.0	29.35	21.0	36.71	37.0	45.37
5.5	29.56	21.5	36.96	37.5	45.67
6.0	29.78	22.0	37.21	38.0	45.96
6.5	29.99	22.5	37.47	38.5	46.26
7.0	30.21	23.0	37.72	39.0	46.56
7.5	30.42	23.5	37.98	39.5	46.85
8.0	30.64	24.0	38.23	40.0	47.15
8.5	30.86	24.5	38.49	40.5	47.46
9.0	31.08	25.0	38.75	41.0	47.76
9.5	31.30	25.5	39.01	41.5	48.06
10.0	31.52	26.0	39.27	42.0	48.37
10.5	31.74	26.5	39.53	42.5	48.68
11.0	31.97	27.0	39.80	43.0	48.99
11.5	32.19	27.5	40.06	43.5	49.30
12.0	32.42	28.0	40.33	44.0	49.61
12.5	32.65	28.5	40.60	44.5	49.92
13.0	32.88	29.0	40.87	45.0	50.24
13.5	33.11	29.5	41.14	45.5	50.56
14.0	33.34	30.0	41.41	46.0	50.87

14.5	33.57	30.5	41.69	46.5	51.19
15.0	33.81	31.0	41.96	47.0	51.51
15.5	34.04	31.5	42.24	47.5	51.84
16.0	34.28	32.0	42.52	48.0	52.16
16.5	34,52	32.5	42.80	48.5	52.49

APPENDIX I

THERMODYNAMIC PROPERTIES OF AIR AND WATER

1. Latent Heat of Vaporization (λ)

$$\lambda = [2.501 - (2.361 \times 10^{-3})\,T]$$

Where: λ = latent heat of vaporization [MJ kg^{-1}]; and T = air temperature [°C].

The value of the latent heat varies only slightly over normal temperature ranges. A single value may be taken (for ambient temperature = 20°C): λ = 2.45 MJ kg^{-1}.

2. Atmospheric Pressure (P)

$$P = P_o\,[\{T_{Ko} - a(Z - Z_o)\} \div \{T_{Ko}\}]^{(g/(a.R))}$$

Where: P, atmospheric pressure at elevation z [kPa]

P_o, atmospheric pressure at sea level = 101.3 [kPa]

z, elevation [m]

z_o, elevation at reference level [m]

g, gravitational acceleration = 9.807 [m s^{-2}]

R, specific gas constant == 287 [J kg^{-1} K^{-1}]

a, constant lapse rate for moist air = 0.0065 [K m^{-1}]

T_{Ko}, reference temperature [K] at elevation z_o = 273.16 + T

T, means air temperature for the time period of calculation [°C]

When assuming P_o = 101.3 [kPa] at z_o = 0, and T_{Ko} = 293 [K] for T = 20 [°C], above equation reduces to:

$$P = 101.3[(293 - 0.0065Z)\,(293)]^{5.26}$$

3. Atmospheric Density (ρ)

$$\rho = [1000P] \div [T_{Kv}\,R] = [3.486P] \div [T_{Kv}], \text{ and } T_{Kv} = T_K[1 - 0.378(e_a)/P]^{-1}$$

Where: ρ, atmospheric density [kg m^{-3}]

R, specific gas constant = 287 [J kg^{-1} K^{-1}]

T_{Kv}, virtual temperature [K]

T_K, absolute temperature [K]: T_K = 273.16 + T [°C]

e_a, actual vapor pressure [kPa]

T, mean daily temperature for 24-hour calculation time steps.

For average conditions (e_a in the range 1–5 kPa and P between 80–100 kPa), T_{Kv} can be substituted by: $T_{Kv} \approx 1.01\,(T + 273)$

4. Saturation Vapor Pressure function (e_s)

$e_s = [0.6108]*\exp\{[17.27*T]/[T + 237.3]\}$
Where: e_s, saturation vapor pressure function [kPa]
T, air temperature [°C]

5. Slope Vapor Pressure Curve (Δ)

$\Delta = [4098.\ e°(T)] \div [T + 237.3]^2$
$= 2504\{\exp[(17.27T) \div (T + 237.2)]\} \div [T + 237.3]^2$
Where: Δ, slope vapor pressure curve [kPa C^{-1}]
T, air temperature [°C]
$e^0(T)$, saturation vapor pressure at temperature T [kPa]
In 24-hour calculations, Δ is calculated using mean daily air temperature. In hourly calculations T refers tothe hourly mean, T_{hr}.

6. Psychrometric Constant (γ)

$\gamma = 10^{-3}\ [(C_p.P) \div (\varepsilon.\lambda)] = (0.00163) \times [P \div \lambda]$
Where: γ, psychrometric constant [kPa C^{-1}]
c_p,specific heat of moist air $= 1.013$ [kJ kg^{-10}C^{-1}]
P, atmospheric pressure [kPa]: equations 2 or 4
ε, ratio molecular weight of water vapor/dry air $= 0.622$
λ, latent heat of vaporization [MJ kg^{-1}]

7. Dew Point Temperature (T_{dew})

When data is not available, T_{dew} can be computed from e_a by:
$T_{dew} = [\{116.91 + 237.3 \text{Log}_e(e_a)\} \div \{16.78 - \text{Log}_e(e_a)\}]$
Where: T_{dew}, dew point temperature [°C]
e_a, actual vapor pressure [kPa]
For the case of measurements with the Assmannpsychrometer, T_{dew} can be calculated from:

$$T_{dew} = (112 + 0.9T_{wet})[e_a \div (e^0 T_{wet})]^{0.125} - [112 - 0.1T_{wet}]$$

8. Short Wave Radiation on a Clear-Sky Day (R_{so})

The calculation of R_{so} is required for computing net long wave radiation and for checking calibration of pyranometers andintegrity of R_{so} data. A good approximation for R_{so} for daily and hourly periods is:

$$R_{so} = (0.75 + 2 \times 10^{-5}z)R_a$$

Where: z, station elevation [m]
R_a, extraterrestrial radiation [MJ m^{-2} d^{-1}]
Equation is valid for station elevations less than 6000 m having low air turbidity. The equation was developed by linearizing Beer's radiation extinction law as a function of station elevation and assuming that the average angle of the sun above the horizon is about 50°.

For areas of high turbidity caused by pollution or airborne dust or for regions where the sun angle is significantly less than 50° so that the path length of radiation through the atmosphere is increased, an adoption of Beer's law can be employed where P is used to represent atmospheric mass:

$$R_{so} = (R_a) \exp[(-0.0018P) \div (K_t \sin(\Phi))]$$

Where: K_t, turbidity coefficient, $0 < K_t \leq 1.0$ where $K_t = 1.0$ for clean air and $K_t = 1.0$ for extremely turbid, dusty or polluted air.

P, atmospheric pressure [kPa]

Φ, angle of the sun above the horizon [rad]

R_a, extraterrestrial radiation [MJ m^{-2} d^{-1}]

For hourly or shorter periods, Φ is calculated as:

$\sin \Phi = \sin \varphi \sin \delta + \cos\varphi\cos\delta\cos\omega$

Where: φ, latitude [rad]

δ, solar declination [rad] (Eq.(24) in Chapter 3)

ω, solar time angle at midpoint of hourly or shorter period [rad]

For 24-hour periods, the mean daily sun angle, weighted according to R_a, can be approximated as:

$$\sin(\Phi_{24}) = \sin[0.85 + 0.3 \; \varphi \; \sin\{(2\pi J/365) - 1.39\} - 0.42 \; \varphi^2]$$

Where: Φ_{24}, average Φ during the daylight period, weighted according to R_a [rad]

φ, latitude [rad]

J, day in the year

The Φ_{24} variable is used to represent the average sun angle during daylight hours and has been weighted to represent integrated 24-hour transmission effects on 24-hour R_{so} by the atmosphere. Φ_{24} should be limited to ≥ 0. In some situations, the estimation for R_{so} can be improved by modifying to consider the effects of water vapor on short wave absorption, so that: $R_{so} = (K_B + K_D) R_a$ where:

$$K_B = 0.98\exp[\{(-0.00146P) \div (K_t \sin \Phi)\} - 0.091\{w/\sin \Phi\}^{0.25}]$$

Where: K_B, the clearness index for direct beam radiation

K_D, the corresponding index for diffuse beam radiation

$K_D = 0.35 - 0.33 \; K_B$ for $K_B \geq 0.15$

$K_D = 0.18 + 0.82 \; K_B$ for $K_B < 0.15$

R_a, extraterrestrial radiation [MJ m^{-2} d^{-1}]

K_t, turbidity coefficient, $0 < K_t \leq 1.0$ where $K_t = 1.0$ for clean air and $K_t = 1.0$ for extremely turbid, dusty or polluted air.

P, atmospheric pressure [kPa]

Φ, angle of the sun above the horizon [rad]

W, perceptible water in the atmosphere [mm] $= 0.14 \; e_a \; P + 2.1$

e_a, actual vapor pressure [kPa]

P, atmospheric pressure [kPa]

APPENDIX J

PSYCHROMETRIC CHART AT SEA LEVEL.

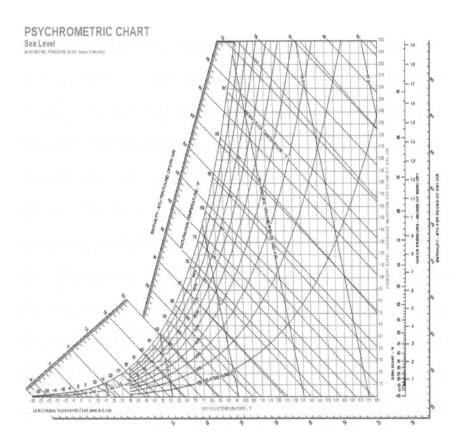

INDEX

A

accumulation, 36, 44, 217
acidity, 230
actual vapor pressure, 23, 24, 26
adjustable outlet system, 46
agricultural economy, 39
agricultural waste recycling industry, 190
agriculture, 49, 106, 130, 175, 193–195, 216, 218, 242
air temperature, 3, 4, 6, 8–10, 23, 119
albedo, 23
alfalfa, 3, 9, 10, 12, 13, 22, 26, 28, 30, 31, 111, 120
   Medicago sativa, 111
   wind speed coefficients, 10
algae, 49
alkathene pipes, 49
alkathene tubes, 49, 102
Allium cepa, 109
almond, 39, 55, 72, 216
American Meteorological Society, 10
American Society of Civil Engineers, 4, 8
American Society of Civil Engineers Task Committee, 22
analysis of variance, 217, 245
apparent density, 52, 102
apple, 39, 45, 55, 58, 71, 72, 219
application efficiency, 45, 91, 102, 120, 242
apricot, 39, 55, 72, 233
aquatic crop, 122
arenosols, 270
arid region, 30, 31, 54, 56, 242
ARS researchers, 3
ASCE committee, 5, 11
asparagus, 134
atmospheric pressure, 26
automatic filtration system, 139
   automation equipment, 139
available soil moisture depletion, 43, 54, 102
available water, 130, 162, 169
avocado nurseries, 39, 71

B

banana, 39, 55, 62, 72, 74, 75, 85, 133, 154, 156, 162–164, 166–171
   Robusta banana, 62
   with conventional fertilizers, 156
   with water soluble fertilizers, 156
bare soil, 10, 111
basal crop coefficient, 13, 14
basics of drip irrigation, 39–45
   description of drip irrigation system, 41
   drip irrigation versus soil salinity, 44
   effects of drip irrigation on crop water use, 40
   effects of irrigation methods, 44
   irrigation intervals, 43
   potential transpiration and consumptive use, 40, 41
bedding plants, 39
bell pepper, 117
benefit cost ratio, 81, 161
beryllium atom, 52
biwall drip tubing, 45, 103
black nightshade, 119
   Solanum nigrum, 119
blackberries, 134
Blaney-Criddle (B-C) method, 2, 3, 5
blueberries, 134
bottle guard, 59, 103
Bowen ratio, 8, 9, 15
Brassica oleracea, 109, 115
Bubbler systems, 138
bubbling patterns, 138
bulbs, 39
Bureau of Plant Industry, 4
button dripper, 47, 103

C

C.N.A.B.R.L. localized drip irrigation, 48
cabbage, 71, 109, 122, 161
   Brassica oleracea, 109
Cache La Poudre river valley, 4
Calcium hypochlorite, 57

Milton Keynes UK
Ingram Content Group UK Ltd.
UKHW031142141024
449569UK00024B/1138